The Sandars Lectures
1899 and 1904

The Printers, Stationers and Bookbinders of Westminster and London from 1476 to 1535

By

E. GORDON DUFF

ARNO PRESS

A New York Times Company
New York 1977

First published London 1906
Reissued 1971 by
Benjamin Blom, Inc.
Reprint Edition 1977 by Arno Press Inc.
LC 77-90066
ISBN 0-405-09135-4
Manufactured in the United States of America

TO

FRANCIS JOHN HENRY JENKINSON

IN REMEMBRANCE OF MUCH KINDNESS

AND MUCH PLEASANT WORK

IN CAMBRIDGE

PREFACE

THE lectures contained in the present volume I delivered as Sandars Reader in two series, the first in the Lent Term 1899, the second in the May Term 1904 and they are thus separated by an interval of five years. Of the first series a small edition was privately printed for presentation, but never published, the second series is now printed for the first time. Though the second part forms a continuation of the first, the two are quite distinct. With the close of the fifteenth century many important changes took place in the English book-trade, and its conditions altered to a great extent, so that the period from 1476 to 1500 has many essential points of difference from the period between 1501 and 1535 and they can with advantage be treated separately. I have thought it best therefore not to attempt to combine the two series, but to issue them, with certain corrections and additions in their original form.

It remains to express my sincere thanks to Dr Jenkinson, the University Librarian who was kind enough to read the proofs, and to Mr H. G. Aldis to whom I am indebted for the excellent index.

E. G. D.

April, 1906.

LIST OF PLATES

(*These plates are the full size of the originals.*)

1. TITLE-PAGE TO W. DE WORDE'S EDITION OF THE
   'BOOK OF GOOD MANNERS' .   .   . *to face* p. 36
   From the unique copy in the University Library, Cambridge.

2. MACHLINIA BORDER USED BY R. PYNSON IN
   'APHTHONII SOPHISTAE PRAEEXERCITAMENTA'
   to face p. 56
   From the copy in the University Library, Cambridge.

3. PAGE OF THE SARUM BREVIARY PRINTED BY R. DE
   NOVIMAGIO AT VENICE IN 1483 .   . *to face* p. 74
   From the unique copy in the Bibliothéque Nationale.

4. BINDING OF FREDERICK EGMONT ON A COPY OF
   'LAVACRUM CONSCIENTIAE' PRINTED AT ROUEN
   to face p. 114
   From the example in the library of Caius College, Cambridge.

5. UNRECORDED DEVICE USED BY W. DE WORDE IN
   THE 'MANIPULUS CURATORUM' OF 1502 . *to face* p. 132
   From the unique copy in the Bodleian Library.

6. TITLE-PAGE TO R. REDMAN'S EDITION OF LYNDEWODE'S
   'CONSTITUTIONS' OF 1534 .   .   .   . *to face* p. 176
   From the copy in the University Library, Cambridge.

7. DEVICE OF CHRISTOPHER ENDHOVEN FROM THE SARUM
   'PROCESSIONALE' PRINTED AT ANTWERP IN 1525
   to face p. 222
   From the copy in the Bodleian Library.

PART I.
1476—1500.

LECTURE I.

THE PRINTERS AT WESTMINSTER.

WHILE the history of the invention and introduction of the art of printing into the various countries of Europe is not only obscure, but still the subject of endless controversy, the history of its introduction into England is now practically settled.

There are no troublesome and incomprehensible documents as in the case of France. No questionable references or undatable fragments such as Dutch and German bibliographers have to contend with. The only attempt that has been made to bring forward an earlier printer than William Caxton is founded upon the misprinted date in the first book printed at Oxford.

In 1664, while the Company of Stationers and the King were quarrelling over the question which had or should have the most power in matters pertaining to printing, a certain Richard Atkyns put forth a tract, now exceedingly rare, called *The Original and Growth of Printing*. In this tract, intended to uphold the

King's rights, attention was drawn for the first time to the Oxford book. "A book came into my hands," writes Atkyns, "printed at Oxford, A.D. 1468, which was three years before any of the recited authors would allow it to be in England." Around this book Atkyns wove a wonderful romance, in the style of the earlier legends about Coster and Gutenberg. Rumours of the new art, he suggests, having reached England, trusted men were sent over to bribe or kidnap an eligible printer and bring him over secretly, along with a press, type, and other *impedimenta*, to England. This was accordingly done, and a certain Frederick Corsellis was conveyed into England, and set up a press in Oxford. One curious point has escaped all commentators on this story, and that is that a real person named Corsellis did come over to England from the Low Countries about that time, and was an ancestor of several well-known London families in Atkyns's time, such as the Van Ackers, the Wittewronges and the Middletons.

Atkyns referred for evidence to documents which have never been found, and his story has met with the disbelief it deserved, but the Oxford book with the date of 1468 not only exists, but still has supporters who consider, or say they consider, the date to be genuine.

Singer in the early part of the century wrote a book in favour of its authenticity, though, as he afterwards attempted to suppress his work, we may conclude he had changed his opinion. Mr Madan of the Bodleian, in his recent admirable history of Oxford printing, clings hesitatingly to 1468, "but quaere" as he would himself say. Generally, however it is agreed that the date is a misprint for 1478. The book has printed signatures, which are not known to have been used before 1472, and when the book is placed alongside the two others issued from the same press in 1479 and printed in the same type, it falls naturally into its proper place, taking just the small precedence which its slightly lesser excellence of workmanship warrants.

Having now disposed of Caxton's only rival, let us turn to Caxton himself. It would, I think, be out of place here to recapitulate however shortly the history of Caxton's early life, since it has been so fully and excellently done in that standard book Blades's *Life of Caxton*. What is more to our purpose is to pass on to the time when, as an influential and prosperous man, he laid the foundations of his career as a printer. By 1463 Caxton had been appointed to the office of governor of the English nation in the Low Countries, a post of considerable importance, and entailing the supervision of trade and traders, and this office he held until about the year 1469. At this latter date he was also in the service of the Duchess of Burgundy, though in what capacity is not stated; but he certainly employed himself at her request in making translations of romances. The *Recuyell of the Historyes of Troye*, a well-known romance of the period, was translated between the years 1469 and 1471, and presented to the duchess in September of the latter year. In the prologue of the printed edition Caxton explains that after the duchess had received her copy, many other persons desired copies also, but that finding the labour of writing too wearisome for him, and not expeditious enough for his friends, he had "practised and learnt, at his great charge and expense, to ordain the book in print, to the end that every man might have them at once."

Now in 1471, when Caxton finished his translation of the *Recueil*, he was living at Cologne, a city remarkable even at that time for the number of its printers, and the first town that Caxton had visited where the art was practised. He had just finished the tedious copying of a large manuscript, so that the advantages of printing would be manifest to him; and we may be tolerably certain that it was about this time and at this town that he took his first lessons in the art and mastered the mechanical processes.

Printing by this time had ceased to be a secret art, nor was there such a demand for books as to make it a very valuable one. The printed books of Germany had at an early date found their way to Bruges, and people's eyes were accustomed to the sight of the printed page, though the nobles still preferred manuscripts, as being more ornamental and costly. There are copies in the Cambridge University Library and at Lambeth of the *Cicero de officiis*, printed at Mainz by Schoiffer in 1466, which were bought in 1467 at Bruges by John Russell, afterwards Bishop of Lincoln, when abroad on a diplomatic mission; and a speech of his, delivered at Ghent in 1470 on the occasion of the investiture of the Duke of Burgundy with the order of the Garter, was one of Caxton's earliest printed productions.

A very strong piece of evidence to my mind that Caxton learnt at Cologne is to be found in the epilogue to the English translation of the *De proprietatibus rerum*, by Bartholomæus Anglicus, which was printed by W. de Worde, Caxton's apprentice and successor, in

1496. This epilogue, written by De Worde himself, contains these lines:—

> And also of your charyte call to remembraunce,
> The soule of William Caxton, first prynter of this boke,
> In Laten tonge at Coleyn, hymself to avaunce,
> That every well disposyd man, may theron loke.

Now this is a perfectly clear statement that Caxton printed a *Bartholomœus* in Latin at Cologne, and we know an edition of the book manifestly printed at Cologne about the time Caxton was there. The type in which it is printed greatly resembles that of some other Cologne printers, and it seems to be connected with some of Caxton's Bruges types. At any rate, the story cannot be put aside as without foundation. It is not, of course, suggested that Caxton printed the book by himself or owned the materials, but only that he assisted in its production. He was learning the art of printing in the office where this book was being prepared, and his practical knowledge was acquired by assisting to print it.

Returning to Bruges, he set about turning his knowledge to account, and in partnership with a writer of manuscripts, named Colard Mansion, began to make or obtain the necessary materials.

Between the years 1471, when Caxton had learned the art at Cologne, and 1474, when he set about obtaining material, printing-presses had started work at Utrecht, Alost, and Louvain. Caxton would most naturally turn for assistance to a town in his own neighbourhood, and there is very little doubt that this town was Louvain, and that the printer who assisted him was John Veldener.

About 1475 their first book was issued, the *Recuyell of the Historyes of Troye*, the first book printed in the English language. It is a small thick folio of 352 leaves, and though not uncommon in an imperfect condition, is of the very greatest rarity when perfect. Two other books were printed by 1476, *The Game and Playe of the Chesse* and the *Quatre derrenières choses*, the latter a very rare book, of which only two copies are known.

In 1476 Caxton obtained a new fount of type, and leaving the first fount with Colard Mansion, who continued to use it for a short time, prepared to set out with his new material for England.

It must have been early in 1476 that Caxton returned and set to work. He took up his residence in Westminster at a house with the heraldic sign of the "Red Pale," which was situated in the Almonry, a place close to the Abbey where alms were distributed to the poor, and where Margaret, Countess of Richmond, the mother of Henry VII., and a great patroness of learning, built alms-houses. The exact position of Caxton's house is not known, but it was probably on some part of the ground lately covered by the Westminster Aquarium.

The first dated book printed in England was the *Dictes or Sayengis of the Philosophres*, translated from the French by Earl Rivers, a friend and patron of Caxton, and edited by Caxton himself, who added the chapter "concernyng wymmen," a chapter which, with its prologue, exhibits a considerable amount of humour.

It is interesting to notice that, as the book is in English, we alone of European nations started our press with a book in the vernacular.

The ordinary copies of the *Dictes* are without

colophon, though the printer and year are in the epilogue, but a copy formerly in the Althorp Library and now at Manchester has an imprint which states that the book was finished on the 18th November, 1477. Although we count the *Dictes or Sayengis* as the first book printed in England on account of its being the first dated book, it is quite possible that some may have preceded it. Between the time of Caxton's arrival in 1476 and the end of 1478 about twenty-one books were printed, and only two have imprints, so that the rest are merely ranged conjecturally by the evidence of type or other details. Now in 1510 W. de Worde issued an edition of *King Apolyn of Tyre*, translated from the French by one of his assistants, Robert Copland, who in his preface writes as follows: "My worshipful master Wynken de Worde, having a little book of an ancient history of a kyng, sometyme reigning in the countree of Thyre called Appolyn, concernynge his malfortunes and peryllous aduentures right espouuentables, bryefly compyled and pyteous for to here, the which boke I Robert Coplande have me applyed for to translate out of the Frensshe language into our maternal Englysshe tongue at the exhortacion of my forsayd mayster, accordynge dyrectly to myn auctor, gladly followynge the trace of my mayster Caxton, begynnynge with small storyes and pamfletes and so to other." Now this Robert Copland was spoken of a little later as the oldest printer in England, so that he may well have known a good deal about the beginning of Caxton's career. We find a very similar case in Scotland. Printing was introduced there mainly for the purpose of printing the Aberdeen *Breviary*, but the first thing the printers did was to issue a series of small poetical pieces by Dunbar, Chaucer, and others, an exactly similar kind of set to the small Caxton pieces in the Cambridge University Library.

In connexion with these Caxton pieces I noticed the other day a strange statement. The writer was speaking of Henry Bradshaw's knowledge of Caxton, and went on to say that "to his bibliographical genius the Cambridge University Library owes the possession of its many unique Caxtons and unique Caxton fragments." The library, however, owes them mainly to the much-maligned John Bagford, who collected the early English books which came to the University with Bishop Moore's library. The monstrous collection of title-pages in the British Museum, generally associated with Bagford's name, was made by the venerated founder of English bibliography, Joseph Ames.

Before the end of 1478 Caxton had printed about twenty-one books. Of these sixteen were small works, all containing less than fifty leaves; of the others the most important is the first edition of the *Canterbury Tales,* of which there is, I think, no perfect copy. Blades speaks of a fine perfect copy in the library of Merton College, Oxford, and remarks also that Dibdin ignorantly spoke of it as imperfect. In Dibdin's time, however, it certainly was imperfect, for I have seen some notes of Lord Spencer's referring to his having sent some leaves from an imperfect copy to the college to assist them in perfecting their own, a courtesy which they repaid by presenting to the library at Althorp their duplicate and only other known copy of *Wednesday's Fast* printed by W. de Worde in 1532.

Among the other books of the period of special interest is the *Propositio Johannis Russell*, which has often been ascribed to the Bruges press, as the speech of which it consists was delivered in the Low Countries. Lord Spencer's copy had a curious history. It is bound up in a volume of English and Latin MSS., and in the Brand sale in 1807 the volume appeared among the MSS., with a note, " A work on Theology and Religion with five leaves at the end, a very great curiosity, very early printed on wooden blocks, or type." It was bought by the Marquis of Blandford for forty-five shillings, and at his sale ten years after cost Lord Spencer £126.

Another interesting book is the *Infancia Salvatoris*, of which the only known copy is at Göttingen, being one of the two unique Caxtons which are in foreign libraries. It was originally in the Harleian Library, which was sold entire to Osborne the bookseller, and was bought with many other books for the Göttingen University. It is in its old red Harleian binding, with Osborne's price, fifteen shillings marked inside, and the note of the Göttingen librarian: "aus dem Katalogen Thomas Osborne in London 12 Maii 1749 (No. 4179) erkauft."

In the first group of books comes also the only printed edition of the *Sarum Ordinale* or *Pica*, which was superseded by Clement Maydeston's *Directorium Sacerdotum*. Unfortunately the book is only known from some fragments rescued from a binding and now in the British Museum. To it refers the curious little advertisement put out by Caxton, the only example of a printer's advertisement in England in the fifteenth century, though we know of many foreign specimens:

"If it plese ony man spirituel or temporel to bye ony pyes of two and thre comemoracions of salisburi use, enpryntid after the forme of this present lettre whiche ben wel and truly correct, late hym come to westmonester in to the almonesrye at the reed pale and he shal haue them good chepe." So far the advertisement; below it is the appeal to the public, "Supplico stet cedula." It seems curious that this should be in Latin, for one would naturally suppose that the ones most likely to tear down the advertisement would be the persons ignorant of that language.

Two copies of this advertisement are known, one in the Bodleian, and another, formerly in the Althorp collection, at Manchester. It has been suggested that both copies may have been at one time extracted from some old binding in the Cambridge University Library. The example at Manchester certainly belonged at one time to Richard Farmer, who was University Librarian, but the Bodleian example was found by Francis Douce in a binding in his own collection.

The group of eight small books in the University Library which I spoke of as perhaps printed earlier than the *Dictes or Sayengis* were originally all bound together in one volume in old calf, and lettered "Old poetry printed by Caxton." This precious volume contained the *Stans Puer ad Mensam*, the *Parvus Catho*, *The Chorle and the Bird*, *The Horse the Shepe and the Goose*, *The Temple of Glas*, *The Temple of Brass*, *The Book of Courtesy*, and *Anelida and Arcyte*, and of five of these no other copies are known.

About 1478-9 was issued the *Rhetorica Nova* of Laurentius of Savona, of which two copies are known, one in the Library of Corpus Christi College, Cambridge, the other in the University Library of Upsala. Now although this book had been known and examined by many for two hundred years, and is printed in the most widely used of Caxton's types, yet it was not recognised as a Caxton until it was examined by Henry Bradshaw in 1861. The colophon says that the work was compiled in the University of Cambridge in 1478, and it was in consequence described by all the early writers as the first book printed at Cambridge. Strype wrote an account of the Corpus copy to Bagford, who in his turn wrote of it to Tanner, and he in his turn communicated it to Ames. Ames then inserted it at the head of his list of books printed at Cambridge, and the mistake, as is usual in such cases, was copied in turn by each succeeding writer on printing.

In 1480 considerable changes are to be found in Caxton's methods of work, owing no doubt to competition, for in this year a press was started in London by a certain John Lettou. He appears to have been a practised printer, and his work is certainly better than Caxton's, his type much smaller and neater, and the page more regularly printed. He also introduced into England the use of signatures. Signatures are the small letters printed at the foot of the page which were intended to serve as a guide to the bookbinder in gathering up the sheets. From the earliest times they were added in writing both to manuscripts and the earliest printed books, but about 1472 printers began to print them in in type, and the habit soon became general. Caxton's use of signatures begins in 1480 and was doubtless copied from the London printer.

At the beginning of 1480 Caxton had printed an indulgence in his large type, the second of his founts, and immediately afterwards the London printer issued another edition in his small neat type. Caxton promptly had another fount cut of small type, and issued with it a third edition of the indulgence.

It is a matter much to be regretted that Henry Bradshaw never issued one of his *Memoranda* on the subject of these indulgences, for he had collected much interesting information, and was the first to point out the variations in the wording of the different issues as well as the discoverer of several unknown examples.

The year 1481 saw the introduction of illustrations, which were first used in the *Mirror of the World*. In it there are two sets of cuts, one depicting various masters, either alone or with several pupils, the other are merely diagrams copied from those found in manuscripts of the work. These diagrams are meagre and difficult to understand, so much so that the printer himself has put several in their wrong places. The explanatory words inside the diagrams, which would no doubt have been printed in type had Caxton had a fount small enough, are written by hand. It is interesting to notice that in all copies of the book the same handwriting is found, though I am afraid it would be unsafe to conclude it to be Caxton's. The period from 1480 to 1483 is the least interesting as regards Caxton's books. Besides the *Mirror of the World* only two books contain woodcuts; the *Catho*, and the second edition of the *Game and Playe of the Chesse*. The two cuts in the *Catho* had been used before in the *Mirror*, but the sixteen in the chess-book are specially cut, though clearly by a different artist from the one who made those for the *Mirror*. Mr Linton in his book on wood-engraving expressed the opinion that many of these cuts were of soft metal, treated in the same manner as a woodblock, but whenever we find any of them in use for a long period, the breaks which occur in them and the occurrence sometimes even of worm-holes show that the cut must have been of wood.

Among the other books of this period are the first and second editions of *Caxton's Chronicle* and *Higden's Polycronicon*. The unique copy of the Latin *Psalter* in the British Museum, a Caxton which remained unidentified until fairly recently, also belongs to about 1480, but perhaps the most interesting book of all is the first edition in English of *Reynard the Fox*. This was translated by Caxton from the Dutch, the translation being finished in June, 1481, and the book evidently printed at once. It is curious that this book, which would lend itself so readily to illustration, was not printed with woodcuts, but Caxton after using them in 1481 made no further move in this direction until 1484, when another group of illustrated books appeared. It always looks as though Caxton, and indeed his own words tend to prove it, was much more interested in the literary side of his work than in the mechanical, and therefore only called in the aid of the wood-engraver when he thought it absolutely necessary. He wished his books to be purchased on their merits alone, and therefore did not try, like the later printers, to use illustrations merely to attract the unwary purchaser. On the other hand, as none of the other printers in

England issued illustrated books, he had no competition to contend with.

A book which may have been printed about this time, but if so has entirely disappeared, is a translation of the *Metamorphoses of Ovid*. In the Pepysian Library is a MS. of books x.-xv., with the following colophon:

Translated and fynysshed by me William Caxton at Westmestre, the 22 day of Apryll, the yere of our lord 1480 And the 20 yere of the Regne of Kyng Edward the fourth." It seems very improbable Caxton would have taken the trouble to make this translation had he not intended it to be printed, and he mentions it in one of his prologues amongst a series of books which he had translated and printed. This MS. was bought by Pepys at an auction in 1688.

Another interesting point to be noticed about it is that it contains the autograph of Lord Lumley who inherited the library formed by the Earls of Arundel. Now William Fitzalan, Earl of Arundel, was one of Caxton's patrons, and the manuscript may have been presented to him by Caxton himself.

The period from 1483 to 1486 is more interesting. First in order comes the first edition of Mirk's *Liber Festivalis* and its supplement the *Quattuor Sermones*. The next is a small quarto pamphlet known as the *Sex quam elegantissimæ epistolæ*, and consisting of letters that passed between Sixtus IV. and the Venetian Republic. The only copy known was found bound up in a volume of seventeenth century theological tracts in the library at Halberstadt, and was sold in 1890 to the British Museum for £200. After these come a series of English writers, Lidgate's *Life of our Lady*;

Chaucer's *Canterbury Tales, Troilus and Cressida,* and *Hous of Fame;* Gower's *Confessio amantis,* and the *Life of St Wenefrede.* The *Canterbury Tales* is the second edition published by Caxton, and has a peculiarly interesting preface by the printer, in which he tells us that having some six years before printed the *Canterbury Tales,* which were sold to many and divers gentlemen, one of the number had complained that the text was corrupt. He said, however, that his father had a very fine MS. of the poem which he valued highly, but that he thought he might be able to borrow it. Caxton at once promised that if this could be done, he would reprint the book. This second edition is ornamented with a series of cuts of the different characters, and one of all the pilgrims seated together at supper at an immense round table. This cut does duty several times later on as the frontispiece of Lidgate's *Assembly of the Gods.*

In the same year as the *Canterbury Tales* appeared two other illustrated books, the *Fables of Esop,* and the *Golden Legend.* The *Esop* has one large full-page cut of Esop used as a frontispiece and which is found only in the copy at Windsor Castle, and no less than a hundred and eighty-five smaller cuts, the work of two if not three engravers, one being evidently the man who made the cuts for the chess-book.

The *Golden Legend* is the largest book ever printed by Caxton. It contains 449 leaves, and is printed on a much larger-sized paper than he ever used elsewhere, the full sheet measuring about two feet by sixteen inches. The frontispiece is a large woodcut representing the saints in glory, while in addition there are eighteen

large and fifty-two small cuts, the large series including
one of the device of the Earl of Arundel, to whom the
book is dedicated. The three dated books of 1485 are
all especially important. The first is the first edition
of the *Morte d'Arthur*, surely the most covetable of all
Caxton's books. For many years only one copy was
known in the library of Osterley Park, and many were
the attempts made by the two great Caxton collectors
in the early years of last century, Lord Spencer and his
nephew, the Duke of Devonshire, to obtain the treasure.
The Duke of Devonshire almost succeeded, but was
foiled by some awkward clause in a deed. However,
another copy appeared at a sale in Wales, wanting
eleven leaves, but otherwise in beautiful condition, and
this was bought by Lord Spencer. The Osterley Park
copy was sold in 1885 for £1950, and went to America,
and after several changes of ownership is now in the
fine library of Mr Hoe of New York. The other two
dated books are the *Life of that Noble and Christian
Prince, Charles the Great*, and the *History of the
Knight Paris and Fair Vienne*. Both of these books
were translated by Caxton from the French. Only one
copy of each is known, and both are in the British
Museum.

After 1485 Caxton's energy began to decline, or at any
rate we know of fewer books having been issued during
the period from 1486 to 1489. The *Speculum vitæ
Christi* and the *Royal Book* belong to 1486, and are
illustrated with woodcuts of a very much superior
execution to those which had been previously in use;
they are not large, but are simply and gracefully
designed. Besides the regular series in the *Speculum*

specially cut for it, a few very small and rather roughly designed cuts are found, evidently cut for use in one of the editions of the Sarum *Horae*, which were issued at an earlier date, but of which nothing now remains but a few odd leaves. It is interesting to notice that in neither edition of the *Speculum* which he printed did Caxton use the full series of the cuts which had been engraved for it; for, several years afterwards, one or two cuts occur in books printed by Caxton's successor, evidently part of the series, and which he had never used himself. To this time may be ascribed the newest Caxton discovery, two fragments printed on vellum of an edition of the *Donatus Melior*, revised by Mancinellus, which were discovered some few years ago by Proctor in the binding of a book in the library of New College, Oxford.

In 1487 Caxton was anxious to issue an edition of the Sarum *Missal*, and, not considering his own type suitable for the purpose, commissioned a Paris printer named William Maynyal to print one for him. Who this Paris printer was is a matter of mystery. In 1489 and 1490 he printed two service books for the use of the Church of Chartres, but is not otherwise known. A George Maynyal, probably a relation, printed at Paris about 1480, and M. Claudin conjectures on somewhat vague grounds, that both were English. The *Missal* is a very handsome book, printed in red and black, with two fine woodcuts at the Canon. The only known copy which belongs to Lord Newton of Lyme Park, appears to have met at an early date with bad treatment, and wants some seventeen leaves, mostly at the beginning.

In this book for the first time Caxton uses his well-known device, consisting of his trade or merchant's mark, with his initials on either side.

Whether this device was cut in England or abroad has long been a vexed question, but as it has no resemblance to any foreign device of the period, and as the execution is poor and coarse, we may conclude safely that it is of native work. Caxton, no doubt, wished to call attention to the fact, which might have escaped notice, that the book was produced for him and at his cost; and so when the copies of the book had been delivered to him at Westminster he had the device cut, and stamped it on the last leaf of each copy. In this edition the portion of the marriage service in English has been omitted by the printer, who has left blank spaces for it to be filled in with the pen. There was an edition of the Sarum *Legenda* issued about the same time, which is known now only from a few odd leaves rescued from book-bindings. It agrees in every way typographically with the *Missal*, it is in the same type, has the same number of lines to the page, every detail the same, so I think we have good reason for supposing that it also was printed by Maynyal for Caxton. Bradshaw suggested Higman, the Paris printer, as the printer of these fragments, so that Maynyal may have had some business connexion with him.

The second edition of the *Golden Legend* came out shortly after this, that is about 1488, and is a difficult book to explain typographically. About 200 leaves are of the first edition, while the beginning, a small piece of the middle, and the end are of the second. Now it is curious that no copy in existence seems to be correctly made up with the full number of second edition leaves, and the most probable explanation seems to be that part of the stock happening to get damaged, a reprint was made to complete what was left, and that sheets were picked indiscriminately. The most nearly perfect second issue that I have seen is the one at Aberdeen, but it is imperfect at beginning and end. The copy in the Hunterian Museum at Glasgow has a second edition ending, and also part of the first quire of the second issue.

In 1489 two editions of an indulgence from Joannes de Gigliis were issued, printed in a type used nowhere else by Caxton and not mentioned by Blades. The earliest noticed of these indulgences was discovered in the following manner. Cotton, who found it at Dublin, published an account of it in the second series of his *Typographical Gazetteer* in 1862, and he there described it as a product of the early Oxford press. Bradshaw obtained a photograph of it, and at once conjectured from the form and appearance of the type that it was printed by Caxton. He immediately communicated his discovery to Blades, who, however, refused to accept it as a product of Caxton's press without further proof, and it was never mentioned in any edition of his books on that printer. The necessary proof was soon afterwards forthcoming, for Bradshaw found that in a book printed by W. de Worde in 1494, the sidenotes were in this identical type, and as De Worde was the inheritor of all Caxton's material, this fount must have belonged to him.

About the same year were issued two unique books, *The History of Blanchardin and Eglantine*, and the *Four Sons of Aymom*.

The *Blanchardin* is unfortunately imperfect, wanting all the end, and it is impossible to say of how much this consisted. The *Four Sons of Aymon* is also imperfect, wanting a few leaves at the beginning. Both books were formerly in the Spencer Library. The *Doctrinal of Sapience* published in 1489 is a translation by Caxton from a French version, and one particular copy of it in the Royal library at Windsor is worthy of special notice. It is printed throughout upon vellum, a material which Caxton hardly ever used, the only other complete book so printed being a copy of the *Speculum Vitæ Christi* in the British Museum. This particular copy of the *Doctrinal* has also a special chapter added "Of the negligences happyng in the masse and of the remedyes" which is not found in any other copy. That it was specially printed is evident from its concluding words, "This chapitre to fore I durst not sette in the boke by cause it is not conuenyent ne aparteynyng that euery layman sholde knowe it."

During the last year or two of his life most of the books issued by Caxton were of a religious nature. Some would have us believe that this was owing to illness or a premonition of his own approaching end, some to the fact that his wife, if the Maud Caxton who was buried in 1490 was his wife, was just dead. Both these ideas seem to me rather fanciful. He no doubt printed what was most in demand. One book issued about this time was certainly not religious. It is a free paraphrase of some portions of the *Æneid* and was translated by Caxton from the French. It does not pretend to be a translation of the original, but was abused soundly by Gavin Douglas, who issued a transla-

classed as one book. There is one cut in the second treatise taken from the *Speculum* series, but no other illustrations.

Caxton used during his career eight founts of type, of which six only are included in Blades's enumeration. The late French type which appeared about 1490-91, and is found in a few of the latest books, such as the *Ars Moriendi* and the *Fifteen Oes*, Blades considered not to have been used until after Caxton's death; and the type of the 1489 indulgences was not mentioned at all. Blades's arrangement, too, of the books under their types, though correct in a certain way, is a very misleading one, for he takes the types in their order, and then arranges all the books under the type in which the body of the book is printed. Now this leads to considerable confusion when different types were in use together. For instance, Caxton started at Westminster with types 2 and 3, and both are used in his first book, but Blades puts the books in type 3 after all those in type 2, and thus the Sarum *Ordinale*, perhaps the second book printed in England, certainly one of the earliest, comes thirty-sixth on his list. Now, though Blades's arrangement was not a chronological one, most writers have made the mistake of thinking so, and have followed it as such, as may be seen, for instance, in the list appended to Caxton's life in the *Dictionary of National Biography*, which follows Blades's arrangement without any reference to his system or mention of the types.

Caxton printed in England ninety-six separate books, and, counting in the three printed by him at Bruges, and the Sarum *Missal*, altogether one hundred, of

tion in 1553, for its many inaccuracies. Amongst the religious books I may mention the *Ars Moriendi*, a little quarto of eight leaves, which was discovered by Henry Bradshaw in a volume of tracts in the Bodleian, and of which no other copy is known, and the very interesting *Commemoratio lamentationis beate Marie*, which is in the University Library at Ghent and which is one of the two unique Caxtons on the Continent. It was, I believe, picked up by one of the librarians bound in a volume of tracts and by him presented to the University Library. This Caxton bought for a trifle in Belgium may be considered as the real successor to the imaginary one picked off the stall in Holland by the celebrated Snuffy Davy of the *Antiquary*.

The *Fifteen Oes* is another of these religious books. Its name is taken from the fact that each of the fifteen prayers of which it is composed begins with O, and it was printed as a supplement to a Sarum *Horæ*, with later editions of which it was generally incorporated. It contains a beautiful woodcut of the Crucifixion, and is also the only existing book printed by Caxton which had borders round the pages. That a *Horæ* to accompany it was printed is most probable, for the Crucifixion is only one of a set of cuts which was used, together with the borders, in an edition printed about 1494.

Though most of the books at this time can only be arranged conjecturally it is probable that the last book printed by Caxton was the *Book of Divers Ghostly Matters*. It consists really of three tracts, each separately printed, the *Seven Points of True Love*, the *Twelve Profits of Tribulation*, and the *Rule of St Benet*; but as they are always found bound together, they are

which ninety-four are mentioned by Blades. It is true that Blades describes ninety-nine books, but he includes two certainly printed at Bruges after Caxton had left, and three printed by De Worde after Caxton's death. But it is not the mere number of the books he printed that makes Caxton's career so remarkable, but the fact that he edited almost every book he issued, and translated a large number. He himself says that he had translated twenty-two, and the statement was made at a time previous to his making several others, and when we consider that amongst his translations is to be included such a large book as the *Golden Legend*, we can only wonder that he printed as much as he did.

Of the exact date of his death we have no evidence, but it evidently must have taken place in 1491. It is unfortunate, too, that no copy of his will has been preserved; for the collection of documents in Westminster Abbey, where it might, with most probability, have been expected to be found, has been searched in vain. The will, besides the interesting personal details which it might supply, would most likely give some information about those engaged with him in business, the assistants who worked his presses, or the stationers who sold his books.

Of his family we know next to nothing. We know that he was married and had a daughter named Elizabeth, who was married to a merchant named Gerard Croppe, from whom she obtained a deed of separation in 1496. Had Caxton had a son he would probably have continued the printing business. As it was the printing materials were inherited by his assistant or apprentice Wynkyn de Worde, who continued to carry on work in his old master's house at Westminster. In his letters of denization, taken out so late as the 20th April, 1496, he is described as a printer, and a native of the Duchy of Lorraine. His name, De Worde, which some have fallen into the mistake of deriving from the town of Woerden in Holland, is clearly taken from the town of Wörth in Alsace; indeed, the printer sometimes uses the form Worth in place of Worde. Although he inherited Caxton's business, which was no doubt a flourishing one, he seems to have started on his own account with very little vigour or enterprise. Indeed, so torpid was the press at that time that foreign printers found it worth their while to produce and import reprints of Caxton's books for sale in this country, books to which I shall refer more fully in a future lecture. We soon see that we have to deal now with a man who was merely a mechanic, and who was quite unable to fill the place of Caxton either as an editor or a translator, one who preferred to issue small popular books of a kind to attract the general public, rather than the class of book which had hitherto been published from Caxton's house.

For the first two years De Worde contented himself with using Caxton's old types, of which he appears to have possessed at least five founts, and in that time he printed five books, the *Book of Courtesy*, the *Treatise of Love*, the *Chastising of God's Children*, the *Life of St Katherine*, and a third edition of the *Golden Legend*. Why this book should have been so often printed is rather a mystery, for, while Caxton issued two editions and De Worde another two before 1500, at the end of the century a considerable number of Caxton's edition still remained for sale at the price of thirteen shillings and fourpence, not a large sum for those days and considering the size of the book.

The *Book of Courtesy*, which is known only from two leaves in the Douce collection at Oxford, was a reprint from Caxton's edition, of which the only known copy is in the Cambridge University Library. In the waste leaves in the Bodleian, De Worde's device is printed upside-down, and for this reason perhaps the sheet was rejected and used to line a binding, and thus preserved for us. The *Treatise of Love* was printed for the translator, whose name unfortunately does not appear, but the translation is dated 1493, and the printing is clearly of the same year. The *Chastising of God's Children*, a deplorably dull book, is interesting typographically as being the first book printed at Westminster with a title-page. Why Caxton never introduced this improvement it is hard to say, for he must have seen many books in which they were used, and a book with one was printed at London before his death.

The imprint of the *Golden Legend* is curious, for though it is dated 1493 it contains Caxton's name. De Worde seems to have reprinted from an earlier edition, merely altering the date, or perhaps he meant the words "By me William Caxton" to refer to the translator rather than the printer.

In 1493, very nearly at the close of the year, De Worde's first type makes its appearance in an edition of the *Liber Festivalis*, the second or companion part of the book, the *Quattuor Sermones*, coming out early in 1494. The type has a strong French appearance, though it retains several characteristics and even a few identical letters of Caxton's founts. It is curious that up to this time De Worde had not put his name to any book, though most of them contain his first device, a copy on a small scale of Caxton's, and evidently cut in metal.

In 1494 two important books were issued, the *Scala perfeccionis* of Walter Hylton, a Carthusian monk, and a reprint of the *Speculum Vitæ Christi*, both being in the late French type of Caxton. The *Scala perfeccionis* is a rare book when it contains the last part, which is only found in two or three copies. It has on the title-page a woodcut of the Virgin and Child under a canopy, and below this the sentence beginning "Sit dulce nomen domini nostri Jesu Christi benedictum," but the engraver in cutting the block has not attempted to cut the words properly, but merely to give their general appearance, so that the result though decorative is almost impossible to decipher.

The *Speculum* of this year has many points of interest, the chief perhaps being that Caxton's small type No. 7 is found in it, the only time it is used in a printed book, though it had been used before in 1489 for printing indulgences. The text of the book is in Caxton's French type, but the sidenotes are in this small Caxton type up to about the middle of the book, whence the notes are continued in the same type as the text. Up till a year or two ago only one copy of this book was known, in Lord Leicester's library at Holkham, but lately another copy, imperfect and in bad condition, turned up amongst some rubbish in the offices of a solicitor at Birkenhead, and is now in the Rylands

Library at Manchester. Three editions of the *Horae ad usum Sarum*, two in quarto and one in octavo, printed in the same type as the other two books, may also be ascribed to 1494. The two in quarto are evidently reprinted from the last edition of Caxton's of which the little treatise called the *Fifteen Oes* formed part, for they have the same borders, and the woodcuts are clearly of sets which belonged to Caxton. The octavo edition is quite different, having no borders, and the woodcuts so far as is known, for the book is only known from a fragment, belong to a set which do not appear to have been used again.

The most famous of the cuts used at this time is one of the Crucifixion formerly used by Caxton in the *Fifteen Oes*, of which a facsimile is given by Dibdin in the second volume of his *Typographical Antiquities*, page 79. He erroneously remarks about it in another place, "The woodcut of the Crucifixion was never introduced by Caxton, it is too spirited and elegant to harmonise with anything that he ever published." It was used frequently after this time by De Worde, and affords us towards the end of the century one of the most useful date-tests for undated books. Between May, 1497, and January, 1498, part of the cap of the soldier who stands on the right of the cross was broken away, so that any book containing this cut with the cap entire must be before 1498. In 1499 the cut began to split, and in 1500 it split right across. Towards the end of 1500 one of the two border lines at top and bottom was cut away. Of course there are for De Worde's books many date-tests, and when they can be worked in various ways and in conjunction, the result may be taken as very fairly accurate. If it were only possible to get once together all the scattered undated books for comparison, they could easily be arranged in their exact order.

In 1495 appeared the *Vitas Patrum*, "the moste vertuouse hystorye of the deuoute and right renowned lyves of holy faders lyvynge in deserte, worthy of remembraunce to all wel dysposed persones, whiche hath be translated out of Frenche into Englisshe by Wylliam Caxton of Westmynstre, late deed, and fynysshed at the laste daye of his lyff." The delay in the bringing out of this work may be due to the large number of illustrations, for it is profusely illustrated; the cuts, however, are very rudely designed and engraved.

In the Pepysian Library at Cambridge is a unique edition of the *Introductorium linguæ latinæ*, edited very likely by Horman, which has the words in the preface, "Nos sumus in anno salutis Millesimo quadringentesimo nonagesimo quinto (1495)," which I certainly take to be the year of printing, especially as another edition of the same book in the Bodleian, also unique, has the last word of the date, quinto, altered to nono, and must have been printed before July, 1499. The small tracts printed from 1495 to 1497 are very difficult to date with any precision, but there are a few of particular interest which may be ascribed to that period, such books, for instance, as the *Information for Pilgrims to the Holy Land*, a work well worth reading for amusement, which cannot be said of many of these books; Fitzjames's *Sermo die lune in ebdomada Pasche*, the *Sermo pro episcopo puerorum*, the *Mirror of Consolation*, and the *Three Kings of Cologne*.

1496 is the year usually ascribed to the edition of Trevisa's translation of the *De proprietatibus rerum* of Bartholomæus Anglicus, and I quoted earlier four lines of verse saying that Caxton had printed the book in Latin at Cologne. The three last lines of the same stanza referring to another matter are also very interesting. Having spoken of Caxton it continues:—

And John Tate the yonger joye mote he broke
Whiche late hathe in Englond doo make this paper thynne
That now in our englyssh this boke is prynted inne.

The watermark of this paper is an eight-pointed star in a circle. The supply of this paper does not appear to have been kept up for long, for I have only found it in two other English books. The *Bartholomæus* contains some very good woodcuts, finer than others of the period, and the press-work seems rather more regular than usual, so that perhaps we may accept the statement of Dibdin that "Of all the books printed in this country in the fifteenth century, the present one is the most curious and elaborate, and probably the most beautiful for its typographical execution." It is only fair to say, however, that the copy described by Dibdin was a very exceptional one. In 1496 also came out a reprint of the well-known *Book of St Alban's*, as it is generally called, a treatise on hunting, hawking, and heraldry, with the addition in this issue of the delightful chapter on fishing with an angle, our earliest printed treatise on the art. There is a woodcut of the angler at the beginning, and we see him busily at work with a large tub beside him, just like the German fisher of to-day, into which he may put his fish and keep them alive.

This book would naturally appeal especially to the richer class, and De Worde not only took especial pains with it, but struck off copies upon vellum, some of which have come down to our own day. From a typographical point of view the book is of great interest, for it is printed throughout in a foreign type which made its appearance in England on this occasion only. It was used at Gouda by Govaert van Os, but he seems to have discarded it about 1490 when he removed to Copenhagen. Besides acquiring this fount De Worde also obtained a number of woodcut capital letters, which are used in all his earliest books, and one or two woodcuts, which he used frequently until they were broken and worn out. It has always been a puzzle to me why, if De Worde had had this fount of type beside him for several years, he never used it before, and why, having used it this once, he never used it again. Not a single letter ever appears in another book, and yet the type is a handsome one.

1498 saw the issue of three fine folios: the *Morte d'Arthur*, of which the only known copy is in the John Rylands Library at Manchester, the *Golden Legend*, of which the only known perfect copy is in the same library, and lastly the *Canterbury Tales* of Chaucer. The only perfect copy of this book was sold lately in the Ashburnham sale for £1000, and is now in a private library in America. The first of these three books, the *Morte d'Arthur*, is a reprint of Caxton's edition, but it differs from it in having illustrations. These are no doubt of native workmanship, and might be justly described as the worst ever put into an English book, being coarsely drawn, badly designed, and incompetently engraved. The *Golden Legend* is a mere reprint of the earlier editions, but is interesting for two points in the colophon. The first is an example of the carelessness of the printers. The words in the earlier editions run, "Thus endeth the legend named in latin legenda aurea, that is to say in Englysshe the golden legende, for lyke as golde passeth all other metals, so this legende exceedeth all other books, wherein be contained all the high and great feasts of our lord" and so on. In this edition a line has been omitted, and the words run, "For like as golde passeth all other metalles, wherein ben contained all the highe and grete festes of our lord." Now although the omission makes nonsense of the whole sentence, it is reprinted exactly the same in the later editions issued by De Worde and Julian Notary.

The other point is the date in the colophon, which runs, "Fynysshed at Westmynster, the viii day of Janeuer, the yere of oure lorde Thousande . cccc. lxxxxviii. And in the xiii year of the reygne of kynge Henry the VII." Now as the 13th year of Henry VII ran from August 22, 1497, to August 21, 1498, it is clear that De Worde in speaking of January 8, 1498, meant 1498 as we would calculate, and not 1499, and therefore that he began his year on the first of January and not on the 25th of March, a most important point to be settled in arranging dated books. Another later proof as to De Worde's dating may be mentioned. In the tracts which he printed between January 1 and March 25, 1509, he speaks of himself as printer to the king's mother, but after Henry VIII succeeded in 1509 he styles himself printer to the king's grandmother, so that he clearly used our method of dating.

About the year 1498, De Worde introduced his second device, the largest of the three used in the fifteenth century. It is almost square, with a broad border, and having Caxton's mark and initials above a flowering plant. Between July and December, 1499, a series of small nicks was cut all round the outside edge, and this gives us a useful clue to checking the dates of several books.

In 1499, De Worde brought out an edition of *Mandeville's Travels*. It was not the first edition published in England by a year or two, but it was the first with illustrations, and most realistic illustrations they are. No doubt it was a very popular book, and the two copies known, one in the Cambridge University Library, the other at Stonyhurst, are both imperfect. Fortunately by means of the two we can obtain an exact collation. This year seems to have been a very busy one. While the dated books in the other years of the fifteenth century never rise above four, in this year there are ten, and a considerable number of undated books can be assigned to this year as well. Among them are a number of small poetical pieces by Lidgate, reprints of Caxton's editions. One of these reprints shows how careless a printer W. de Worde was. He reprints the *Horse, the Shepe, and the Ghoos*, from a copy of Caxton's wanting a leaf, but never noticing anything wrong prints straight ahead, making of course nonsense of the whole.

All De Worde's quarto tracts were got up in the same style, the title at the top of the first leaf printed in one of Caxton's types, below this a woodcut not always very apposite to the subject of the work. There were two stock cuts of masters with large birches and their pupils seated before them, one of these being among the material obtained from Govaert van Os. These of course were suitable for grammars and school-books. Caxton's cuts for the Sarum *Horae,* the Crucifixion, The tree of Jesse, the three rioters and three skeletons, the rich man and Lazarus, and David and Bathsheba, came in very useful for theological books. The only special cut, that is, one specially cut for the particular book and not belonging to a series, that I have found, is that on the title of the *Rote or mirror of consolation,* which depicts seven persons kneeling before an altar, above which two angels hold a monstrance.

At the end of the year 1500, De Worde moved from Westminster into Fleet Street at the sign of the Sun, the earliest book from the new address being dated May, 1501. This from the point of view of the bibliographer was an extremely well-timed move, for we can at once put all books with the Westminster imprint as before 1501, and all with the London one after 1500, thus dividing clearly the fifteenth and sixteenth century books. At the time of his moving he seems to have got rid of a considerable portion of his stock; some seems to have been destroyed and some sold, for many cuts which had belonged to De Worde or to Caxton are found afterwards in books printed by Julian Notary. De Worde seems to have been a successful business man, for when he moved into Fleet Street he occupied two houses close to St Bride's Church, one his dwelling-house and the other a printing-office, for which he paid the very high tithe rent of sixty-six shillings and eightpence.

The number of books printed by Wynkyn de Worde in the fifteenth century, counting in different editions of the same book, is 110, and of a considerable number of these only a single copy is known. It would seem probable that the printer, when issuing a small book, printed only a small number of copies, preferring to set up the type for a new edition rather than burden himself with much unsaleable stock. And it is curious how these various editions have been accidentally preserved. Only two copies are known of a book called the *Rote or mirror of consolation*, printed by De Worde in the fifteenth century, one of them is in the Pepysian library, the other in Durham Cathedral. Yet these two are of quite different editions, the one at Durham being certainly about 1496, the other certainly after the middle of 1499. Of the *Three Kings of Cologne* we have two editions, though only three copies are known. Indeed, for some time it was thought that each copy represented a different edition, as the copy in the British Museum, evidently bound up separately out of a volume of tracts, had had the last page of the tract preceding it bound in in place of the correct title-page.

Looking at the very large number of small books which De Worde printed between the end of 1496 and 1500, it is surprising how many are known from single copies. I have kept for many years a register of all the copies of early English books which are to be found anywhere, and taking the quartos printed by W. de Worde, which number altogether 70, I find that out of that number 47, that is more than two-thirds, are known to us now from single copies or fragments. And I feel certain that we owe the preservation of the majority of these to a cause we are now doing our best to destroy. A few worthy people centuries ago made collections of these tracts and bound them up in immensely stout volumes, which gave them an air of importance in themselves, and tended to preserve the tracts inside in a much better manner than if bound separately. I do not think I am exaggerating when I say that a hundred and fifty of the rarest that De Worde printed during his whole life would have been found a hundred years or so ago bound up in about twelve volumes. Some twenty-two of the rarest of W. de Worde's in the Heber Library came to him in one volume. Thirteen unique tracts which sold at the Roxburghe sale for £538, were in a single volume when the Duke purchased them fourteen years before for £26. I need only refer you to the University Library, a large number of whose unique Caxton and De Worde tracts came in three or four volumes. Then again, when so many are known only from fragments or single copies we may imagine what a large number have absolutely disappeared.

Some have been lost of late years or have disappeared since they were described. Three unique W de Worde books of the fifteenth century were supposed to have perished in a fire in Wales in 1807 but fortunately they had been sold by the owner of the library a short time before the fire. Others seem to have drifted into libraries whose owners know nothing about them. There is a unique De Worde printed before 1501, entitled the "*Contemplacyon or meditacyon of the shedynge of the blood of our lorde Jhesu Cryste at seven tymes.*" This was seen and described by Herbert, who very likely saw it when it was sold at the Fletewode sale in 1774. Since then we have no record of the book, and though every year more information about private collections is published I can come upon no trace of it.

Beside the genuine books which have disappeared, by this I mean books which have been described by a trustworthy bibliographer, there are others which may reasonably be supposed to have existed, and one clue to these is afforded by the woodcuts. W. de Worde for example had certain series of cuts, specially made for certain books, but when he wished to decorate the title-page of a small tract, which was not itself to be otherwise illustrated, he used an odd cut out of his sets. Now when we can trace in different tracts odd cuts manifestly belonging to a series, we may reasonably suppose that the book for which the series was engraved must have been printed.

To give a couple of instances. In the unique copy of Legrand's *Book of good manners* in the University Library without date, but printed about the middle of 1498, are two cuts, which really belong to a series made to illustrate the *Seven wise masters of Rome*. These cuts are fairly accurate copies of those used by Gerard Leeu in his edition of 1490. At a considerably later date De Worde did issue an edition of the *Seven wise masters*, illustrated with the series of which the two mentioned above formed part, and showing at that time marks of wear. Now as De Worde had the series cut by the beginning of 1498, I think it most probable that an edition of the book was then issued, for it is unlikely that he would go to the trouble of cutting the set unless he was preparing to print the book.

TITLE-PAGE TO W. DE WORDE'S EDITION OF THE 'BOOK OF GOOD MANNERS.'

Again, before the end of the fifteenth century De Worde had a series to illustrate *Reynard the Fox*. One cut is found on the first leaf of an edition of Lidgate's *The Horse, the sheep, and the goose*, in the University Library, another on the title-page of Skelton's *Bowge of Court* in the Advocates' Library at Edinburgh. In the collection of the University Librarian is a fragment of an edition of *Reynard*, evidently printed by W. de Worde about 1515, and this contains a third cut agreeing absolutely in size, in workmanship, and in style with the other two.

In this case again it seems probable that an edition illustrated with these cuts appeared before 1500.

The last press at Westminster during the fifteenth century is that of Julian Notary, which while it started in London about 1496 and only moved to Westminster in 1498, is more suitably taken in this place on account of its connexion with Wynkyn de Worde.

The first book issued was an edition of *Albertus de modis significandi*, printed in a neat Gothic type, but containing no information in its colophon beyond that it was printed in London at St Thomas the Apostle's, probably close to the church of that name, and not at a house with that sign. There is also a printer's mark containing three sets of initials, I. N. for Julian Notary, I. B. for Jean Barbier, and I. H. for someone unidentified, but who there are some reasons for supposing to have been Jean Huvin, a printer at Rouen, who was associated in the production of books for the English market.

In 1497 the same printers issued an edition of the *Horae ad usum Sarum*, very neatly printed, and with delicate borders round the pages. All that remains of the book is a fragment of four leaves, rescued from a book-binding, but this luckily contains the colophon, telling us that it was printed at St Thomas the Apostle's, for W. de Worde. This book also contains the device with the three sets of initials.

In 1498 appeared a Sarum *Missal*, the first edition printed in England, and though otherwise well got up, the musical parts have the drawback of being without notes, only the staves having been printed, though whether this was done by design or merely because the printers had no musical type remains unknown. From the colophon of the *Missal* we learn that the printers, Julian Notary and Jean Barbier, had settled at Westminster, and had printed the book at the command and expense of W. de Worde. On the last leaf is Caxton's device, and on the title-page that of the printers. Of this book five copies are known, and of the four I have examined, the copy in the University Library is the only perfect one. About the fifth, belonging to the Duke of Sutherland, I have no information.

I. H. it is clear had left the firm, and though the printers use the same device as before, the initials I. H. have been cut out of it.

In 1499 Jean Barbier also disappeared, for in the edition of the *Liber Festivalis* and *Quattuor Sermones* which appeared in that year the printer's mark has again been altered. All initials have been cut out and the name Julianus Notarii inserted in type. This form of the name suggests that he was not a notary as is generally stated, but the son of one. I have never been able to see a perfect copy of this book though Herbert describes one which he said was in the Inner Temple Library, but my inquiries there met with no success. Hain in his *Repertorium Bibliographicum* mentions a copy which seems not to be the one noticed by Herbert.

In April, 1500, Notary printed a most minute edition of the *Horae ad usum Sarum*, it is in 64s as regards folding, and a printed page measures an inch and a quarter by an inch. Only a fragment of it is known, a quarter sheet containing sixteen leaves, but that luckily contains the colophon. It was very likely copied from another edition of the same size, which was printed at Paris the year before, but this point cannot be determined, as the only copy of the latter which existed was burnt with the greater part of the Offor collection. All we know now of it is the meagre note in the auctioneer's catalogue, "imperfect, but has end with imprint"—and he has not given the imprint!

The colophon of Notary's *Horae* tells us that it was printed in King Street, Westminster. King Street is the short street at the bottom of Whitehall in a straight line between Westminster Abbey and the Foreign Office, though in Notary's time it appears to have extended from Westminster to Charing Cross. Lewis in his life of Caxton, says that Caxton's printing-office was in King Street, but I do not know of any reason for his assertion.

The last of Notary's books printed at Westminster is an edition of Chaucer's *Love and complaintes between Mars and Venus*, with some other pieces. This rare little book, having passed through the collections of Farmer, the Duke of Roxburghe, Sir Mark Masterman

Sykes and Heber, is now at Britwell. The colophon runs: "Thys inpryntyde in westmoster in kyng strete. For me Julianus Notarii." In spite of the word For, I think the book was printed by Julian Notary himself. It contains two cuts, reversed copies of two of Caxton's.

At what time Notary left Westminster cannot at present be settled, but probably almost immediately after W. de Worde. When his next dated book was issued in 1503 he had moved to London, and with his departure from King Street to Pynson's old house near Temple Bar printing ceased altogether in Westminster.

LECTURE II.

THE PRINTERS AT LONDON.

THE art of printing was introduced into London in 1480, three years after Westminster, two years after Oxford, and probably one year after St Alban's, by a printer called Joannes Lettou. The name evidently denotes that he came originally from Lithuania, of which the word Lettou is an old form. One thing is at once apparent when we come to examine his work, and that is that he was a skilled and practised printer, producing books entirely unlike Caxton's, and bearing every appearance of being the work of a foreign press. Where he learned to print it is impossible to find out, but whence his type was obtained no one can have the least doubt: it was certainly brought from Rome.

The type is identical with that used by a printer at Rome in 1478 and 1479, who really ought to have some connexion with English printing as his name was John Bulle. In his Roman books he describes himself as from Bremen. If it were possible to arrive at any explanation how a man from Bremen could be described as a Lithuanian, I should at once assume John Lettou and John Bulle to be identical, since the one apparently begins where the other leaves off. However, until some reasonable explanation is forthcoming it will be best to consider them as different people.

Lettou seems to have been assisted during the first two years of his career by a certain William Wilcock, but who this man may have been I have not been able to discover, unless he was a certain William Wilcock who is mentioned in the State Papers as having been presented to the living of Llandussell in 1487. The two books printed by Lettou were the *Questiones Antonii Andreae super duodecim libros metaphisice* and the *Expositiones super Psalterium* of Thomas Wallensis. Both these books are printed in a neat small Italian gothic, with two columns to the page and forty-nine or fifty lines in a column. The first of the two books, the *Antonius Andreae*, is a small folio of 106 leaves, and almost all the six known copies are imperfect. The book is very probably reprinted from the edition printed at Vicenza in 1477, the only earlier edition than the present, which like it, is edited by Thomas Penketh. Penketh was a friar of the Augustinian house at Warrington, but went later as teacher of theology to Padua. He returned to Oxford in 1477, where he also taught theology, and was probably living there when his book was being printed at London.

The second of these two books is printed in exactly the same style and form as the first, with the exception of having fifty lines to the column in place of forty-nine. In the imprint the book is ascribed to the " Reverendissimus dominus Valencius," that is Jacobus Perez de Valentia, who was, however, not the author of this work, though he did write a commentary on the Psalms. The real author was a certain Thomas Wallensis or de Walleis. Henry Bradshaw, who discovered the mistake, gives the following explanation of it: "This edition is printed from an incomplete copy, and from the words of the colophon 'Reverendissimi domini Valencii,' the final s having been misread as an i, the work has been confounded with the commentary of Jacobus Perez de Valencia, which was printed at that place in 1484 and 1493. The v for w and the absence of the Christian name would also serve to create the confusion, or at any rate to perpetuate it."

Three editions of the indulgence of John Kendale were printed by Lettou in 1480. The first two have been preserved in a very curious manner. It was a common custom of the early binders to paste a thin strip of vellum down the centre of each quire of paper in order to prevent the thread which ran down the centre of the quire and stitched it to the bands of the binding from cutting through the paper. A copy of a foreign printed Bible, which appears to have been bound in England, perhaps by Lettou himself, and which is now in the library of Jesus College, Cambridge, has the centre of each quire throughout the book lined with a strip of vellum, part of cut up copies of these two indulgences. Indulgences having their year printed upon them soon went out of date, and as they were of vellum and printed only on one side were very much used by bookbinders for lining bindings. These two indulgences were issued early in the year and have the date 1480, but no mention is made of the pontifical year of the pope. The third indulgence, of which a copy is in the British Museum, also dated 1480, has besides, the date of the pontifical year, "the year of our pontificate the tenth," and as the popes dated like the kings, from the exact date of their accession or coronation, this copy must have been printed after August 7, 1480, on which date the tenth year of Sixtus IV began.

After the printing of his two books and editions of the indulgence, Lettou entered into partnership with a printer called Wilhelmus de Machlinia, a native, as his name shows, of Mechlin in Belgium. Together they printed five books, the *Tenores Novelli* of Littelton, the *Abridgement of the Statutes*, and the *Year-books* of the thirty-third, thirty-fifth, and thirty-sixth years of Henry VI.

For these books the printers used a small very cramped black letter, abounding in abbreviations, and often difficult to read. It appears to have been designed after the law hand of the period. The edition of the *Tenures* is the only one of these books with an imprint, and it contains the names of both printers, and the statement that the book was printed in the city of London, "juxta ecclesiam omnium sanctorum." There were, however, several churches in London at this time dedicated to All Saints, and it is not possible now to settle which particular one was meant. Complete sets of these five books are in the British Museum and the Cambridge University Library.

The entire change in the character of the books produced after Machlinia had joined Lettou shows that his strong point was legal printing, and during his continuance in business he seems to have printed all the law-books issued in England. But perhaps the most marked peculiarity of his partnership is the extraordinary deterioration in the books produced. The work of Lettou was marked by excellence of typography and the many improvements introduced by an evidently practised printer. As soon as Machlinia joined him the work became slovenly. It might be supposed that Mr William Wilcock, who had defrayed the expense of Lettou's work, had either tried it as a speculation and found it a poor one, or had only wished the two books to be specially printed for his own use and had then left the printer to shift for himself. It is curious, too, that Lettou's neat type should have entirely disappeared. The real reason for this probably was that though it was very neat it had none of the abbreviations necessary in a type used for printing law-books.

While Lettou remained in the firm the work, though much deteriorated, retained a certain amount of regularity. All the books had signatures and were regular in size, though their appearance was not good. After the issue of these five books Lettou seems to have ceased printing, but the type was used for one more book, which it will be well to notice here, *The History of the Siege of Rhodes*. This was written in Latin by Gulielmus Caorsin, vice-chancellor of the Knights of Malta, and was translated into English by John Kay, who styles himself poet-laureate to Edward IV. It gives an account of the great victory of the Rhodians against the Turks and the death of Mahomet.

It is the only early English printed book which we cannot definitely ascribe to any particular printer. By most early writers it was classed as a production of

Caxton, and Dibdin places it under Caxton in his *Typographical Antiquities*, though he there expresses a doubt as to its being his work. "The typography," he says, "is so rude as to induce me to suppose that the book was not printed by Caxton. The oblique dash for the comma is very coarse; and the adoption of the colon and the period, as well as the comparatively wide distances between the lines, are circumstances which, as they are not to be found in Caxton's acknowleged publications, strongly confirm this supposition." Five years later, writing in the *Bibliotheca Spenceriana*, he seems to have settled more accurately. "I have very little doubt," he writes there, "of its having been executed by Lettou and Machlinia, or by the former of these printers, rather than by Caxton. The letters, however, great and small, especially the larger ones, and some of the compound smaller ones, bear a strong resemblance to the smallest types of our first printer; but on a comparison with those of the *Tenures* of Lyttelton and the *Ancient Abridgement of the Statutes*, printed by Lettou and Machlinia, the resemblance is quite complete." The type is certainly that used by Lettou and Machlinia, and the considerable difference in appearance from the other five books is caused by the text being in English, which makes more difference than would be imagined, and also that there are very few of the abbreviations which crowd the other books. Then again the lines of type are spaced out, giving the page a much lighter appearance.

Though the dedication is to Edward IV, it does not necessarily follow that the book was printed before his death, for the early printers in reprinting a manuscript would keep to the preface as there written. It might, however, have been printed as early as 1483, and immediately the law books had been completed. Who the printer was I do not think can ever be settled. When it was printed Machlinia had probably started by himself with his new types, and I do not think it can have been printed by Lettou, as it has not the signatures to the pages which he invariably used.

We may, I think, date the break up of the partnership of John Lettou and William de Machlinia about 1482–83, and from that date onwards Machlinia worked alone. He seems to have made a fresh start with new type, for he has at least three founts which had not been used before. The difficulties in the way of making any arrangement or arriving at any definite conclusions about his books are very great. We know that he printed at least twenty-two books, and not one single one is dated. Signatures, directors, headlines, seem to be present or absent without rule or reason. There is hardly any method of arranging the books in groups, every book stands alone in splendid isolation.

The only division possible is according to the type used in the books, and in this way we can separate them into two groups. Those of the first group are printed in two founts of a square gothic type, and as in colophons of the two books in this series which possess them the printer speaks of himself as living near the "Flete-bridge," we call these books the ones printed in the Fleet-bridge type. The other group are in a regular English type, similar in general appearance to some of Caxton's or that used by the printer of St Alban's, and in the imprint to one of these books

Machlinia speaks of himself as printing in Holborn, so that we speak of this series of books as printed in the Holborn type. As Flete-bridge was at the east end of Fleet Street and a considerable distance from Holborn it is impossible that these two addresses could apply to one office.

It is probable that the Fleet-bridge group is the earlier, so we will take it first. In it there are altogether eight books. Three folios, the *Tenures of Lyttelton*, the *Nova Statuta*, and the *Promise of Matrimony*, four quartos, the *Vulgaria Terentii*, the *Revelation of St Nicholas to a Monk of Evesham*, and two books by Albertus Magnus, the *Liber Aggregationis seu de secretis naturae*, and the *Secreta mulierum*, and one small book, probably a 16°, an edition of the *Horae ad usum Sarum*.

The two books of Albertus Magnus are certainly the most neatly printed, the press work being tidy and regular, which was not generally the case with this printer's productions.

The copy of the *Secreta mulierum* in the University Library is an interesting one, though, unfortunately, imperfect. On the first leaf which is blank there is a certain amount of writing, amongst other things the following sentence: "Annus domini nunc est 1485 in anno Ricardi tercii 3°." This note, supposing it to have been written at the time to which it refers, and there is no reason to doubt it, must have been written between June 26 and August 22, 1485, showing that at any rate the book was printed before that date. The other book of Albertus Magnus, the *Liber aggregationis*, has a colophon stating that it was printed by " William de Machlinia in the most wealthy city of London, near the bridge vulgarly called the Flete-bridge." The wealth of London seems to have impressed the alien printer, for he always applies the word "*opulentissima*" to that city.

The small *Horae* we have little information about, for we know of its existence only from nineteen leaves scattered about the country. There are eight in Corpus Christi College, Oxford, seven in the British Museum, four in Lincoln Minster, and two in the University Library, Cambridge. These have all been extracted from bindings, and in the cases where we know the particular bindings from which they came these bindings were the work of the same man, whose initials were W. G. From the way in which the leaves were printed, and the way in which they were afterwards folded, a point too technical and difficult of description to touch on here, we may pretty safely say that the *Horae* was a 16° and not an 8°. It may be worth while remarking that the early printers used only the simple folding, which with each successive folding exactly halves the size of the previous one. The sheet folded into two leaves produced folio size, this folded again once made 4to, folded again 8°, again 16°, again 32°, and again 64°. The duodecimo or 12°, which depends on more complicated folding, was quite unknown.

The *Horae*, so far as we can see from what remains, contained no illustrations, but it had an engraved border which was used round the pages beginning certain portions of the book. This engraved border we afterwards find in Pynson's hands, and it is the only definite

D.

link connecting him with this press. Bradshaw, in his paper on the "Image of Pity," suggests that Ames, who quoted this book in his *Typographical Antiquities*, had seen a complete copy, but as he describes it merely as "a book of devotions on vellum" and adds no particulars I think that he simply described it from the few leaves in his own possession, which are now in the British Museum in the great so-called Bagford volumes of despoiled title-pages.

The *Revelation of St Nicholas to a Monk of Evesham* is one of the most remarkable volumes of the fifteenth century, very well worth reading, for it is full of early English stories and allusions. (I may say in passing that Mr Arber has issued a cheap reprint of it.) The story tells of a man who was taken through purgatory and was shown various people whom he had known or heard of and listened to their stories. It seems to me very curious that no other editions of the book were issued in early times: it seems exactly the kind of book which must have been popular. Typographically, the book is interesting as showing an excellent example of wrong imposition, that is that when the one side of the sheet had been printed, the other side was put down upon its form of type the wrong way round, and consequently the pages come all in their wrong order, page 1 being printed on the first side of the first leaf, page 14 follows it on the other side, then page 16, then page 4, and so on. Now, most printers who had done this stupid thing, and it was not an uncommon accident, would have destroyed the sheet and reprinted it. Not so Machlinia. He printed off some more copies of the wrong sheet and, cutting it up, pasted the four pages in their proper places. In one of the two known copies this has had unfortunate results, for some curious inquirer, noticing the pages pasted together, has tried to separate them to find out what was underneath, and they have suffered severely in the process.

The *Vulgaria Terentii* is the last of the quartos in this group. It is a book that was often printed, but of the present edition the copy in the University Library is the only one remaining, and it, unfortunately, is slightly imperfect.

Of the folios, the *Nova Statuta* is the most important, and also by far the commonest, for I have examined over a dozen copies myself, and I know of a good many more. The book must have been printed after April, 1483, as the subject-matter runs up to that date, while an action in Chancery relating to it was tried between 1483 and 1485. The *Promise of Matrimony*, another folio in this type, consisting only of four leaves, relates to the agreement made in 1475 between Edward IV and Louis XI for the marriage of the Princess Elizabeth of York and Prince Charles, afterwards Charles VIII, King of France.

I have noticed that in nearly all the copies of the law-books printed by Lettou and Machlinia, or Machlinia alone, that I have examined, the initial letters, which were filled in by hand in colour, appear to have been done by the same person; the letter roughly in red, and a twirl or two by way of ornament in pale green or blue. I suppose the subject of the books was so severely practical that unless this had been done before the book left the office it would never have been done at all. However, in English printing generally, though the spaces were left for fine initials, I can remember very few books with them filled in in any but the plainest way, a contrast to the beautiful work so often found in Italian books.

The last group of books, which number fourteen, are called the Holborn type books, because, in the imprint of one of the two books that contain them we find the words, "Enprente per moy william Maclyn en Holborn." The general type used for the text of these books is very remarkably like that used by Veldener at Utrecht and by John Brito at Bruges, and may, perhaps, have been obtained by Machlinia from abroad, though it is of the same school of type as several used in England.

The most important book issued in this series is an edition of the *Chronicles of England*. It is a very rare book, but there is an imperfect copy in Cambridge in the Barham collection in Pembroke College. The space for the initial letters, as is usually the case with early books, has been left blank to be filled in by the rubricator, but in one copy that I have seen, the initials have been filled in in gold, not gold leaf, but gold paint, and this is the only example of its use that I have found in an early English book. Another curious point about the book is that though it is a folio, a folio of 238 leaves, yet in all copies leaves 59 and 66, the first and last leaves of a quire, are printed on quarto paper. I thought once that perhaps for some reason these leaves had been cancelled and reprinted, but it seems more probable that the printer had for the moment run out of his supply of ordinary-sized paper, and had to use some of a much larger size cut in half. A similar case of mixed papers is found in the *Nova Rhetorica* printed at St Albans in 1480, which is partly quarto and partly octavo.

Three editions were published by Machlinia of the curious *Treatise of the Pestilence* by Canutus or Kamitus, Bishop of Westeraes in Sweden, and of each edition only one copy is known, one in the British Museum, one in the University Library, and one sold lately in the sale of the Ashburnham Library and now at Manchester. I must warn anyone who uses Dibdin's *Typographical Antiquities*, that the facsimile page of the book which he gives is made up from the upper part of the first leaf of the Cambridge copy and the lower part of the same leaf of the Manchester copy, which he must have seen when it was in the possession of Triphook, the bookseller, so that the resulting facsimile is rather puzzling. The fact that one of these editions, that in the British Museum, has a title-page, makes us inclined to put it to rather a late date, but at any rate it is the earliest title-page in an English printed book.

Another book in this group, by far the commonest and best known, is the *Speculum Christiani*, ascribed to a writer named John Watton, a curious medley of theological matter interspersed with pieces of English poetry. The colophon states that the book was printed for and at the expense of a merchant named Henry Vrankenbergh. About this merchant I could find out nothing until, curiously enough, on my last visit to Cambridge a fortnight ago, my attention was drawn by a friend to a note in the Descriptive Catalogue of ancient deeds in the Record Office, where is a note of a " Demise to Henry Frankenbergk and Barnard van Stondo, merchants of printed books, of an alley in

St Clement's Lane, called St Mark's Alley, 10th May, 1482."

This is, I believe, the earliest note relating to foreign stationers or merchants of printed books in England, but I hope from the same source we may expect to obtain many more as soon as the endless series of documents in the Record Office are calendared.

An edition of the *Vulgaria Terentii* was also printed in this type. An almost perfect copy was added to the British Museum Library some years ago, and a considerable portion of another copy is in the library of Caius College.

Machlinia also printed two of the *Nova Festa*, the *Festum visitationis beate Marie virginis* and the *Festum transfigurationis Jesu Christi*. The first of these is only known from two leaves which had been used to line the boards of the binding of Pynson's *Dives and Pauper*. Of the second a beautiful and perfect copy is in the library of the Marquis of Bath. It is curious to notice that it contains not only the feast according to the Sarum use, but also according to the Roman use.

The last three of Machlinia's books to be noticed are the three which, though undated themselves, contain certain evidences of date. The first of these contains the statutes made in the first year of Richard III, and, as this first year ran from June 26, 1483, to June 25, 1484, the book cannot be earlier than the second half of the latter year.

The second book is one about which I am very much inclined to doubt whether it was printed by Machlinia at all, but rather by Veldener, who used apparently identical type; and though I have had for several years under my charge at Manchester the only copy of the book known I cannot make up my mind about it. It is an edition of the *Regulae et ordinationes* of Innocent VIII, and could not at any rate have been printed before the very end of 1484. The type seems newer in appearance than any of Machlinia's, though to all appearance identical. Dibdin, with his usual readiness, helps us by remarking, "It presents us with the same character or general appearance of type as that which Caxton and Machlinia occasionally used. It is not much unlike the St Alban's type."

The last production is a Bull of Innocent VIII confirming the marriage of Henry VII and Elizabeth of York. It was reissued in 1494 by Alexander VI, and there the date is given as 27th March, 1486.

Two copies of this Bull are known, one in the library of the Society of Antiquaries and one in the Rylands Library, Manchester. Both are in poor condition, and show signs of having been used at one time to line the boards of a binding.

Richard Pynson was by birth a native of Normandy, but practically nothing is known of his personal history. It is probable that he was educated at the University of Paris, for we find in a list of students in 1464 the name "Ricardus Pynson Normannus," and this may very well be the printer. It was, however, in Normandy that he learned to print, probably from Guillaume le Talleur, a noted printer of Rouen, as may be seen by certain small habits connected with printing which he fell into, and which are very typical of Rouen work. Although we have only circumstantial evidence, evidence depending on a number of almost trifling details to back up the statement, it seems now almost certain that Pynson succeeded Machlinia. My own impression is that he succeeded immediately on the death or retirement of the latter, with hardly any interval. A very strong reason for this impression is that had any long time elapsed between the cessation of Machlinia's press and the commencement of Pynson's, England would have been left without a printer who could set up law French. Caxton and Wynkyn de Worde were presumably unable to do it, at any rate they printed no books of the kind except some statutes of Henry VII, and it must be remembered that in Henry VII's reign for the first time the statutes were written in English. I do not mean to suggest that Pynson ever worked with Machlinia, but only that when the latter ceased to work Pynson came over and started in his place, perhaps taking over some of his printing material or even starting work in his old office. The engraved border which Machlinia had used in his Sarum *Horae*, the only piece of ornament he seems to have possessed, we find used afterwards by Pynson, and it is a very common thing to find Pynson's earliest bindings lined with waste leaves of Machlinia's printing. Had Pynson worked with Machlinia we should have expected the latter's founts of type to have passed into his hands, as Caxton's were inherited by Wynkyn de Worde, but they did not. Indeed they totally disappeared, and what we do find of Machlinia's in Pynson's hands is merely the refuse that we might expect a printer to find in an office just vacated by another. Had Pynson not been ready to take over the place this waste stuff would have been destroyed. The question

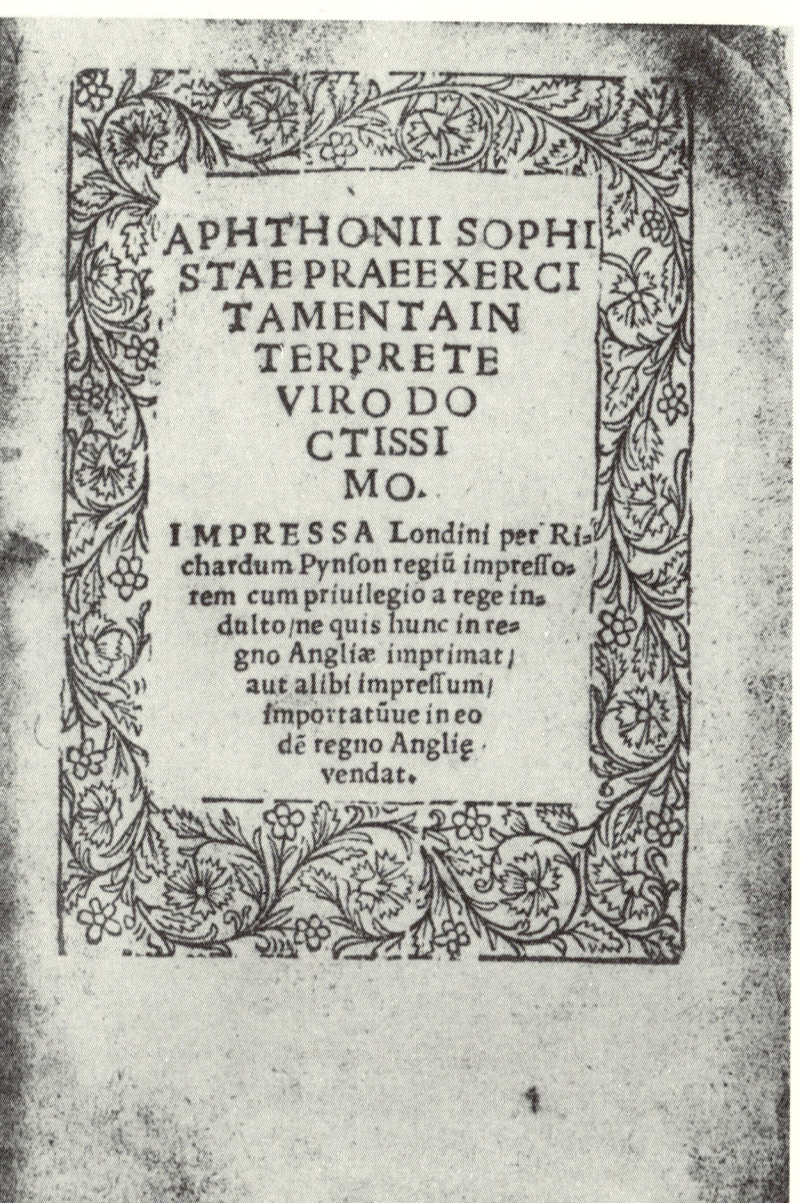

MACHLINIA BORDER USED BY R. PYNSON.

is, then, when did Machlinia cease or Pynson begin? I should say that Machlinia ceased much later than is supposed and Pynson began much earlier, and that the two events happened between 1488 and 1490. At first when Pynson arrived he was without material, so he commissioned Le Talleur at Rouen to print for him the two law-books most in demand, *Littelton's Tenures* and *Statham's Abridgement of Cases to the end of Henry VI.*

Probably in 1490–1491 he began printing on his own account. His first dated book was issued in November, 1492, but five books, if not more, can be placed earlier; these are an edition of Chaucer's *Canterbury Tales*, a *Latin Grammar*, a poetical book, and two or more *Year-books*. The *Chaucer* is a particularly fine book, printed in two sizes of type a larger for the poetry, and a smaller for the prose, which is printed in two columns. It is also illustrated with a set of woodcuts illustrating the different pilgrims. It is interesting to notice that these cuts were altered in some cases while the book was passing through the press in order to serve as the portrait of another pilgrim. The serjeaunt with a little alteration becomes the doctor of physick, the squire becomes the manciple. There has been a good deal of controversy as to the date of the printing of this book and whether it could have appeared before Caxton's death in the latter part of the year 1491. Pynson in his prologue, which is rather confused and difficult to understand, says, speaking of Chaucer: "Of whom I among alle other of his bokes, the boke of the tales of Canterburie, in whiche ben many a noble historie of wisdome policie mirth and gentilnes. And also of vertue and holynes whiche boke diligently ovirsen and duely examined by the pollitike reason and ouirsight of my worshipful master William Caxton accordinge to the entent and effecte of the seide Geffrey Chaucer and by a copy of the seid master Caxton purpos to imprent, by ye grace, ayde, and supporte of almighty god." I think had Caxton been dead Pynson would have alluded to it in some way, speaking of him, perhaps, as " my late worshipful master " or " my worshipful master late dead." The term worshipful master does not imply that Pynson had been an apprentice or assistant to Caxton, but was merely a courteous way of referring to the printer and editor whose work he was about to reprint. Blades in his life of Caxton, speaks of Pynson's having used Caxton's device, but this mistake has arisen through a made-up book in the British Museum, a copy of Bonaventure's *Speculum vite Christi*. The copy wanted the end, and some former owner, in order to make the book look more complete, has added a leaf with Caxton's device printed on it.

The *Latin Grammar* is known only from three leaves, one in the Bodleian and two in the British Museum. The leaf in the Bodleian appears from an inscription upon it to have been used to line a binding as early as 1494. The book was printed entirely in the large black type of the *Chaucer*, the first of Pynson's types, and which he does not appear to have used after 1492.

Another book printed about this time was a book of poetry of a quarto size. All that is at present known of this book are two little strips making part of a leaf, and each containing six lines of verse, three on each side. I found these fragments a year or two ago amongst a bundle of uncatalogued leaves in the Bodleian, but was not able at the time to determine from what book they came. The story is apparently of someone who having been in purgatory is allowed to revisit the world in order to warn others of what he had seen there. This was a common story, and the occurrence of part of a line " But y the goste of guido " makes it certain that the fragments belong to some version of a work called *Spiritus Guidonis*.

The two other books in this series are two year-books, for the first and ninth years of Edward IV. All these early books, with the exception, of course, of the poetry fragments, contain Pynson's first device, which consists of his monogram in white upon a black background, not at all unlike in style that used by Le Talleur at Rouen, with whom he had been associated. When the device was used in November, 1492, a small alteration had been made in it, so that from the state of the device as well as by the type used we are able to settle which books belong to this earliest group.

In 1492 Pynson's first dated book appeared, an edition of the *Doctrinal* of Alexander Grammaticus, editions of which had already been printed abroad in considerable numbers. Pynson's was not copied from any of these, having a different commentary, but who this commentary is by I have not yet been able to ascertain.

This book was only discovered quite lately, and I came upon it by a fortunate accident. The owner, or rather guardian, of it happened to have read in some book that the earliest dated book of Pynson's was issued in 1493. Knowing that he had an earlier one he wrote to the British Museum about it, and I heard casually that the book had been sent to them to examine. I went up to London immediately to see if I could see the book, but was told it had been returned, nor could I obtain any information as to where it was to be found. Luckily, the owner was so far interested as to write a note to one of the papers mentioning the existence of the book, and also the place where it was preserved—the Grammar School at Appleby. The following Saturday I set off to Appleby, and had the pleasure of examining it at my leisure. It is a beautiful copy, quite perfect, in its original binding, and, as one would have hoped, with end leaves taken from Machlinia's *Chronicle*. It has a perfectly clear Latin imprint which runs: "And thus ends the commentary of the Doctrinale of Alexander, printed by me Richard Pynson of the parish of St Clement Danes outside the bar of the new Temple at London the 13th day of the month of November in the year of the incarnation of our Lord 1492." From this colophon it is clear that if Pynson did commence work in Machlinia's old office, which was in Holborn, he had by this time removed to other premises. The commentary in the book is printed in a very small, neat type which Pynson had probably had made for him abroad, as it contained no w. I am sorry that the discovery of this book has thrown out of order the list of Pynson's types which I gave in the introduction to my *Facsimiles of Early English Printing*. In 1903 this volume was sold by the school trustees, and is now in the British Museum.

In 1493 appeared Henry Parker's *Dialogue of Dives and Pauper*, which was always considered, before the discovery of the *Doctrinale*, Pynson's first dated book. It is printed in a new and handsome type, and this is the only dated book in which it is used, though there are four undated quartos in the same type, which may be put down to the same year. These are the *Festum nominis Jesu*, one of the *Nova Festa* printed as supplements to the Sarum *Breviary*, the *Life of St Margaret*, Lidgate's *Churl and Birde*, and an edition apparently of some statutes or a similar work known from two leaves in the library at Lambeth. Of the *Festum nominis Jesu* one copy is known, bound up in a volume with several other tracts, among them being Caxton's *Festum transfigurationis Jesu Christi*. It was for a while in the Congregational Library in London but was eventually sold to the British Museum. Three printed leaves from the beginning of the poem amongst the fragments in the Bodleian are all that remain of the *Life of St Margaret*. The Lidgate's *Churl and Birde* after passing through the sales of Willett, the Marquis of Blandford, Sir F. Freeling and B. H. Bright, passed with the Grenville Library into the British Museum.

Two editions of Mirk's *Liber Festivalis*, each known from a single copy, one in the Pepysian Library, the other in the University Library, belong probably to this time. The various changes in the book are interesting to trace. In the earliest editions there are no references to, or additional chapters for, the new feasts which were then coming into use; then come editions with the extra feasts printed together at the end as a kind of supplement to the book, and finally we get the editions with these extra feasts put into their proper places in the body of the book. The edition in the Pepysian Library is without these extra feasts, while that in the University Library has them as a supplement of ten leaves at the end. In the next edition, which was printed about the end of 1493 by W. de Worde, the feasts have been incorporated into their proper places.

In 1494 Pynson reverted to his earlier types and issued a translation by Lidgate from Boccaccio called the *Falle of Princes*, remarkable for its charming woodcuts. In this book, for the first time, Pynson used his second device, a large woodcut containing his initials on a black shield with a helmet above on which is perched a small bird. This I imagine is meant for a finch, a punning allusion to his name, since pynson is the Norman name for a finch. Round the whole is a border of flowering branches, in which are birds and grotesque beasts. This device supplies us later with a most useful date test, for the edge split in 1496 and the piece broke off entirely towards the end of 1497. After 1498 the use of this device was discontinued but it was not destroyed, and it made a solitary reappearance, in a sadly mutilated state in an edition of a grammar of Whitinton in 1515. Probably in this year (1494) Pynson issued his edition of the *Speculum vitae Christi*, of which an almost perfect copy is in King's library. It is illustrated with a large number of neat woodcuts, which are copied more or less from Caxton's illustrations to the same book, though they are by no means identical with them, as has been often stated. As a general rule Pynson's cuts are of very much better execution and design than either Caxton's or De Worde's, and though not in all cases good, as for instance in the *Canterbury Tales*, yet they never sink to the very bad drawing and engraving so often found in the works of the other two printers.

An edition of the *Hecyra* of Terence bears the date January 20, 1495, but no other dated book of this year is known, and 1496 was hardly better, having only the two grammars, the *Liber Synonymorum* and *Liber Equivocorum*, the latter usually wrongly attributed to Joannes de Garlandia; but many undated books of very considerable interest appeared about this time. Two of these, the *Epitaph of Jasper, Duke of Bedford*, and the *Foundation of Our Lady's Chapel at Walsingham* are to be found in the Pepysian Library. Jasper Tudor, Duke of Bedford, and half-brother of Henry VI, died on December 21, 1495, and the book must have been printed early in 1496. It is ostensibly written by one Smarte, the keeper of the hawks to the Duke, and begins as follows:—

Rydynge al alone with sorrowe sore encombred
In a frosty fornone, faste by Severnes syde
The wordil beholdynge, whereat much I wondred
To se the see and sonne to kepe both tyme and tyde.
The ayre ouer my hede so wonderfully to glyde
And how Saturne by circumference borne is aboute
Whiche thynges to beholde, clerely me notyfyde
One verray god to be, therin to haue no dowte.

The end runs:—

Kynges prynces moste souerayne of renoune
Remembre oure maister that gone is byfore
This worlde is casual, nowe up nowe downe
Wherfore do for your silfe, I can say no more
Honor tibi Deus, gloria et laus
Qd' Smerte maister de ses ouzeaus.

This poem has been attributed to Skelton, though I do not know for what reason. On the title-page is a special cut, not used elsewhere, of Smarte kneeling, with his hawk on his wrist, and presenting with his other hand a book to a person standing. The *Foundation of Our Lady's Chapel at Walsingham* is a small tract relating to the priory of the Augustinian canons of St Mary, once one of the most important places of pilgrimage in England, and which was described by Erasmus. The first leaf, which would have contained the title, is wanting, but the text begins on the second:—

> Of this chapell se here the fundacyon
> Bylded the yere of crystes incarnacyon
> A thousande complete, sixty and one
> The tyme of Sent Edward kyng of this region.

About this time appeared the first English edition of *Mandeviles Travels*, the only edition, I think, issued without illustrations, and a little reprint of Caxton's *Art and Craft to Know Well to Die*, of which the only known copy, formerly Ratcliffe's, is in the Hunterian Museum at Glasgow.

Another poetical tract is the *Life of Petronylla*, beginning:—

> The parfite lyfe to put in remembraunce
> Of a virgyn moost gracious and entere
> Which in all vertu had souereyn suffysaunce
> Called Petronylla petyrs doughter dere.

This little tract consists of four leaves, and though only three copies are known at present it is probable that more are in existence, for the book seems to occur in all the sales of large libraries which have occurred within the last hundred years.

About 1496 Henry Quentell, a Cologne printer, had issued the first edition of the *Expositio Hymnorum et Sequentiarum*, according to the use of Sarum, but it was found that several hymns and sequences were omitted, so Pynson issued two supplements, one of sixteen leaves to the hymns, another of six to the sequences.

Another rather quaint book issued about this time is a kind of vocabulary or phrase book in English and French. "Here is a good book to lerne to speak french, Vecy ung bon liure a apprendre a parler fraunchoys." The book contains also specimens of letters in French relating to trade, in fact it was evidently intended as a manual for people who had business relations with France.

Two more editions of the *Nova Festa*, the *Festum transfigurationis* and the *Festum nominis Jesu* were issued about this time. The only copies known of these two books are in a private library in Somerset, and I have not yet had an opportunity of examining them.

In 1497, or perhaps slightly earlier, Pynson began to use his third device, made probably to take the place of the second which had split, and taking warning from it he had the new one cut in metal. The device consists of the shield and monogram supported by a man and a woman, with the helmet and bird above. The border, which is cut on a separate piece, contains birds and foliage, with the Virgin and Child and a saint in the bottom corners. In the lowest part of the frame a piece in the form of a ribbon has been cut out for the insertion of type. In consequence of the weakness of that particular place the small piece of border below the ribbon began to be pushed inwards, and by 1499 there was a distinct indentation in the border. This got deeper and deeper year by year, until the piece broke off entirely in 1513. The first dated book in which it occurs is the 1497 edition of Alcock's *Mons perfectionis*, but it occurs in several of the undated books that can be placed about 1496.

Between 1495 and the end of 1497 Pynson issued the plays of Terence, the first classic (with the exception of an Oxford edition of the *Pro Milone*, which is known from a few leaves) that had been printed in England. The six plays were evidently issued separately and not as a volume, for they differ considerably typographically. There is some difficulty, too, in determining in what order they were issued.

In 1498 there are seven dated books, one of them being the sermon of Bishop Alcock called *Gallicantus*, and he is so pleased with jesting on his name that he prefaces the text of his sermon with a little black picture of a cockerel, which he also used as a device. Another edition of the *Doctrinale* of Alexander the Grammarian was issued this year, but I have not yet seen the book, as the only copy known belongs to Lord Beauchamp.

In 1499 a very interesting book was printed by Pynson. This was the *Promptorius Puerorum*, a Latin-English dictionary ascribed to a monk of Lynn. The imprint tells us that the book was printed for Frederick Egmondt and Peter post pascha. Frederick Egmondt was an important stationer, and no doubt Peter post pascha was a stationer also, though what name in the vernacular can be represented by post pascha remains an unsolved riddle. Mr Albert Way, in his edition of the *Promptorium parvulorum*, applies a curious amount of misplaced ingenuity to the question of the identity of these two stationers. "We find about the time in question," he says, speaking of the name Egmondt, "a distinguished person of that family, possibly the patron of Pynson, Frederic, son of William IV, Count of Egmond. In 1472 he received from his uncle, the Duke of Gueldres, the lordship of Buren; he was named governor of Utrecht by the Archduke Maximilian in 1492; two years later Buren was raised to a count in reward of his services. There was a Peter, an illegitimate brother of his father, who might have been living at that time; what was his surname does not appear." Another book printed about this year was the *Elegantiarum viginti praecepta*, a book which I am fond of for a peculiar reason. I found once a leaf of it in the Signet Library, Edinburgh, and, not knowing that any copy was in existence, set to work to reconstruct the book from the leaf. I counted the lines, and comparing with foreign editions, conjectured the size and structure of the book, and knowing how Pynson would make a title-page with a woodcut, and the woodcut he would probably use, I made up a description of the book, taking the title from an early bibliography. At last I heard of a perfect copy in a private library which the owner was kind enough to allow me to examine. When the book arrived I found I had not only got the collation right, but by a lucky inspiration had selected the correct woodcut for the title-page. As it happens I might have spared myself the trouble, for I found afterwards a fairly accurate collation of the book in an authority I had not consulted.

A curious prognostication for 1499 is in the Bodleian. It is addressed to Henry VII, and was drawn out by a William Parron, who lived at Piacenza and called himself doctor of medicine and professor of astrology. Another prognostication for 1502, by the same author, was printed by Pynson, and some fragments are in the library of Westminster Abbey. He also wrote an astrological work on Henry, Duke of York, afterwards Henry VIII, in 1502, of which there are manuscripts in the British Museum and Bibliothèque Nationale, but it does not seem ever to have been printed.

The *Missal* which Pynson printed in 1500 is perhaps the finest book printed in the fifteenth century in England. It was produced at the expense of Cardinal Morton, whose arms appear at the beginning, and Pynson has introduced into the borders and initials a rebus on the name consisting of the letters Mor surmounting a barrel or tun. Five copies of this book are known, three being on vellum. One of the latter copies, slightly imperfect, is in the library of Trinity College, Cambridge. In the copy at Manchester all the references to St Thomas and his service, which had been scraped out and erased according to the command of Henry VIII, have been entirely filled in by some pious seventeenth century owner in gold.

In the imprint, after setting forth that the book was printed at the command of Cardinal Morton, Pynson adds the date, January 10, 1500. Now, as Cardinal Morton died on September 15, 1500, I think we have here a strong piece of evidence that Pynson, like De Worde, began his year on January 1. For if he had begun it on March 25, then January 10, 1500, would be after the Cardinal was dead, and Pynson would surely have spoken of him as lately dead, or in some way alluded to the loss of his patron.

The *Book of Cookery*, belonging to the Marquis of Bath, was also printed this year. It begins: "Here beginneth a noble boke of festes royalle and cokery, a boke for a pryncis housholde or any other estates, and the makynge therof accordynge as ye shall fynde more playnly within this boke." Then follows an account of certain great banquets, the feast at the coronation of Henry V; "the earle of Warwick's feast to the king, the feast of my lorde chancellor archbishop of York at his stallacion in York," and so on. After the account of the feasts comes the more practical Calendar of Cookery.

Two editions of the *Informatio Puerorum*, a small grammatical work founded upon the *Donatus*, were issued about this time. In the colophon of one it is stated that the book was printed for George Chasteleyn and John Bars. I have found no reference anywhere to John Bars, but George Chasteleyn was an Oxford bookseller, carrying on business at the Sign of St John the evangelist in that city. It was for him also that in 1506 Pynson printed an edition of the *Principia* of Peregrinus de Lugo. About this time there was no press in Oxford, so that books for use in the schools had to be printed in London.

In a scrap-book in the British Museum are some leaves of an edition of the romance of *Guy of Warwick*.

which may be ascribed to Pynson, and they are printed in a curious mixture of his early types. These leaves were discovered in 1860 in the binding of a copy of Maydeston's *Directorium Sacerdotum*, printed by Pynson in 1501, and an account of them was sent by their discoverer, who signs himself E. F. B., to *Notes and Queries*. Now, of this edition of the *Directorium*, only two copies are known, one in the British Museum and one in Ripon Cathedral, and I should very much like to know from which copy these leaves were obtained.

During all the period from 1490 to 1500 Pynson was busy issuing editions of law-books, more than a quarter of his productions being of this class, and it is probable that a considerable number more printed in the fifteenth century may yet be discovered. They are not of a nature to attract much interest, and are generally very badly catalogued, or catalogued in collections and not separately, and in one great English library at least they have no more detailed press mark than Law Room, so it is needless to say I have not yet examined such books as they may have in that library.

Though he did not print so many books as De Worde in the fifteenth century, nevertheless Pynson was evidently a more enterprising and careful printer. He had seven distinct founts of type, all of which were made for him and not inherited from other printers, and the works he produced were of a much more scholarly nature, though this becomes more apparent in his work during the sixteenth century. His patrons were often learned and distinguished men, for whom he produced such splendid work as the *Morton Missal*, and he became later the recognised king's printer. In the fifteenth century he printed altogether eighty-eight books known to us.

Pynson, like De Worde, very considerately moved to a new address at the end of the century; previous to 1501 he was in St Clement's parish, outside Temple Bar, which was the limit, I think, of the parish, but afterwards moved inside Temple Bar, where he carried on business at the Sign of the George. The colophon to the *Book of Cookery*, printed in 1500, says, "Imprinted without Temple Bar"; the colophon to the *Directorium Sacerdotum* of 1501 says, "intra barram novi templi," so that the date is pretty accurately fixed.

LECTURE III.

THE STATIONERS.

In speaking of the history of the printed book in the fifteenth century I have so far dealt only with the printers of London and Westminster; to-day I propose to touch on the books printed abroad for the English market and the stationers who sold them. In the early days the different businesses of a publisher, a bookseller, and a bookbinder were often carried on by one man, who was called a stationer. He bought books wholesale, sometimes having whole editions specially printed for him, he bound them, and then sold them like an ordinary bookseller. He also probably in England, as was certainly done on the Continent, sent round vans full of books to the various provincial towns, timing his arrival as far as possible to coincide with the local fairs.

A considerable number of the books printed abroad for sale in England have no connexion with any particular stationer, but were probably brought over by an agent of the printer and sold in lots to different stationers.

The earliest book printed abroad definitely for sale in England is the edition of the Sarum *Breviary* printed at Cologne about 1475. Of this book nothing is left but a few leaves, and the imprint, if it possessed one, is not known. Only one other book is known printed in the same type, an edition of the *Homilies*, but it, unfortunately, has no imprint, so that we have no clue as to who may have been the printer. I cannot help thinking that perhaps Caxton may have had something to do with having this book printed, commissioning it either on his own account or for some friend in England, for it is unlikely that a printer in so distant a town would have issued such a book on his own account, and the probable date of its printing coincides more or less with Caxton's departure for England.

In 1483 a book was printed at Venice for sale in England, curiously enough another edition of the Sarum *Breviary*. The copy in the Bibliothèque Nationale, the only one known, is a very beautiful book, printed on vellum and quite perfect. There is a rather painful history attached to it. In 1715 this unique book came to the University Library, Cambridge, as part of the library of Bishop Moore which was presented to the University by George I. In the latter half of the eighteenth century it appears to have been purloined along with a great many other rarities by a certain Dr Combe. It then found its way into the collection of Count Justin MacCarthy, who formed the largest library of books printed on vellum ever brought together by a private collector (he had over 600 of such books), and at his sale in 1815 it was purchased for the Paris Library for fifty-one francs. The printer of the book, Reginaldus de Novimagio, does not appear to have had any connexion with England, nor does the imprint mention for whom the book was produced. It is curious that he should have been chosen as the printer of this *Breviary*, for it seems to have been the only liturgical work he issued, and nothing among his other productions has any connexion with England. Of course English people passed through Venice in large quantities as it was the starting-point for pilgrims to the Holy Land, and many ecclesiastics of high position went on this journey, so that perhaps one of these travellers, seeing the beautiful work done at Venice, and knowing that no printer at home was equal to the task of producing such a book in a fitting manner, commissioned the printing of the *Breviary*. It is sad to think that so beautiful a book has been lost to England through the dishonesty of a reader in the Library. We can only regret that the negotiations between the Duke of Devonshire and the representatives of Count MacCarthy for the purchase of the library *en bloc* fell through, and that the Duke and Lord Spencer, who both bought considerably at the sale, did not secure it, for then at any rate it might have been in England, though not in its proper place.

In the year 1484 some important Acts were passed relating to the trading of foreigners in this country. The ninth chapter ends: " Provided always that this act or any parcel thereof, or any other act made, or to be made in this said parliament, shall not extend, or be in prejudice, disturbance, damage, or impediment, to any artificer, or merchant stranger, of what nation or country he be, or shall be of, for bringing into this

PAGE OF THE SARUM BREVIARY OF 1483.

realm, or selling by retail, or otherwise, any books written or printed, or for inhabiting within this said realm for the same intent, or any scrivener, alluminor, binder or printer of such books which he hath, or shall have to sell by way of merchandise, or for their dwelling within this said realm, for the exercise of the said occupations; this act or any part thereof notwithstanding."

This Act it will be seen, which was not repealed until 1534, gave absolute liberty to foreign printers and stationers to trade and reside in England. That it succeeded in its object of encouraging the immigration of stationers and craftsmen and the importation of books, is clear from the words of the Act of 1534: "Whereas by the provision of a Statute made in the firste yere of the reygne of Kynge Richarde the thirde, it was provided in the same acte that all strangers repayryng into this realme might lawfully bring into the saide realme printed and written bokes to sell at their libertie and pleasure. By force of which provision there hath comen into this realme sithen the makynge of the same, a marvellous number of printed bookes and dayly doth. And the cause of the making of the same provision semeth to be, for that there were but few bookes and fewe printers within this realme at that time, whiche could well exercise and occupie the said science and crafte of printynge. Nevertheless, sithen the making of the saide provision, many of this realme being the Kinges naturall subjectes, have given them so diligently to lerne and exercise the saide craft of printing that at this day there be within this realme a great number of connyng and experte in the said science or crafte of printing, as able to exercise the saide crafte in all pointes as any stranger in any other realme or country."

Though the preamble of this Act speaks only of printing, it was mainly directed against the foreign bookbinders and stationers. By it it was forbidden to import any foreign printed books ready bound, and no one was to buy from any foreigner residing in England any books except "by engrosse," that is, wholesale. This you will see completely stopped the trade of the foreign binder in the English market, and absolutely did away with the foreign stationer in England. One effect of the Act is apparent in the extraordinary number of letters of denisation taken out at that date. In 1582 Christopher Barker wrote: "In the time of King Henry VIII there were but few printers and those of good credit and competent wealth, at whiche time and before there was another sort of men, that were writers, lymners of bookes and dyverse thinges for the Churche and other uses called stacioners; which have and partly to this day do use to buy their bookes in grosse of the said printers, to bynde them up and sell them in their shops, whereby they well mayntayned their families."

The fifty years then between 1484 and 1534 are the really interesting years in the history of the English book trade, when it was free and unprotected, but though we have a fair amount of information about the latter half of this time, the earlier half is almost destitute of any kind of records. The books of the original company of stationers in London have all disappeared, and we are dependent mostly on incidental references in deeds, in wills, or other legal documents.

Two years before the Act was passed, namely in 1482, we know of two foreign booksellers who had come to London, Henry Frankenberg and Bernard van Stondo, who rented an alley in St Clement's Lane called St Mark's Alley. From their names they would appear to have come from the Low Countries, but we know nothing about them or their business beyond the fact that Frankenberg commissioned his fellow-countryman, William de Machlinia, who was printing in London, to print for him an edition of the *Speculum Christiani*, about which I spoke in my last lecture. Their names in the deed and Frankenberg's name in a colophon are the only clues we have to the existence of two probably important booksellers. So also in the very year of the Act we find foreign dealers in books trading in Oxford with the resident university stationer. In 1485 Peter Actors, a native of Savoy trading in London, was appointed by Henry VII, Stationer to the King.

About 1486 at Louvain, Egidius vander Heerstraten printed an edition of the *Regulae Grammaticales* of Nicolas Perott, which contains a great number of passages in English. These are very curious, and seem to have been translated by one not very conversant with the language. Here is a passage which refers to the fifteenth century substitute for compulsory football: "who someuer of my discipulis goyeth awey fyrst from the gammyng wt owt my licence i shal smyte his hande wyt a rode. And yf he do the samyn thyng twyss i shall also beet hym wyt a leyshe." In another place, having translated the Latin phrase, "Quintilianus est eloquens sed nihil ad Ciceronem," "Quintilian is a wel spoken man but nothyng to Tully," he adds another and more personal example: "Helia Perott is fayr but nothing to Penelope."

I am not sure whether we ought to consider this book as one printed for the purpose of exportation to England, or whether it was not rather intended for the use of English students at the foreign universities. This is made more probable from the fact that in a few cases we have words translated into Dutch prefaced by "as we say." I have seen it stated that a similar edition was printed by the same printer with explanations in French, but I have not been able to verify the existence of any copy.

About 1486, too, was issued the first edition of the Sarum *Missal*, printed, it is supposed, at Basle by Wenssler, though some doubts have been raised as to whether it was really printed at Basle on account of the appearance of the music type. It is a very handsome folio volume of 278 leaves, printed in a large Gothic type in red and black. The printer has not attempted to print the English portions of the wedding service, but has left blank spaces where they occur, so that they might be written in by hand. The first few editions of the Sarum *Missal* are all similar in this respect, but it is curious that Caxton, who had an edition specially printed for him, should not have supplied the printer with correct copies for these small portions of the service.

In the next few years a few grammatical books were issued, printed as a rule in the Low Countries. In 1486 Gerard Leeu printed the *Vulgaria Terentii*, a series of Latin sentences with translations into English, an edition reprinted from the Oxford one of a year or two earlier. This book is sometimes found printed as a supplement to the *Grammar* by John Anwykyll, and of this *Grammar* there are two foreign editions, one printed by Paffroed at Deventer in 1489, and another rather later by Henry Quentell at Cologne. The *Grammar* does not contain an author's name, but in the prefatory verses written by Petrus Carmelianus he is referred to as Joannes. There are also verses written to William Waynflete, Bishop of Winchester, and founder of Magdalen College, Oxford, congratulating him on having persuaded this Joannes to edit the *Grammar*. The book is supposed to have been intended for the use of the Magdalen College School, in which the two grammarians John Anwykyll and John Stanbridge were masters, and is supposed to have been the work of Anwykyll. The two earliest editions were printed at Oxford, but by 1486 the Oxford Press had stopped work and the two succeeding editions were printed abroad.

The *Liber Equivocorum* and *Liber Synonymorum*, the former wrongly attributed to Joannes de Garlandia were also printed in the Low Countries, the first at Deventer by Paffroed, the second at Antwerp by Thierry Martens in 1493. The *Liber Synonymorum* has the commentary of Galfridus Anglicus. A copy of this book sold in the Ratcliffe sale in the last century was described as having been printed at Antwerp in 1492, but this must have been, I suppose, a misprint for 1493.

Three more books printed in the Low Countries I

ought to mention before turning to France. One is an edition of Clement Maydeston's *Directorium Sacerdotum*, printed by Gerard Leeu in 1488, of which there is a copy in the Cambridge University Library.

Another is an edition of the Sarum *Horae*, also printed by Leeu, which I am afraid has to be spoken of at present as a lost book. The only fragment known, an unused half sheet containing eight leaves, had been used to line the binding of a copy of the *Scriptores rei rusticae* printed at Reggio in 1496, in Brasenose College library; Bradshaw saw the fragment and took down a description of it, but on its return to Oxford it was mislaid and is not to be found.

The third book is another edition of the Sarum *Breviary*, printed at Louvain in 1499 by Thierry Martens. The only copy known is in the Musée Plantin at Antwerp. Leaving the Low Countries for a time we will turn to France.

The *Missal* printed for Caxton in 1487 I have already described in an earlier lecture, so I can pass on to the edition which succeeded it, that printed by Martin Morin, the celebrated printer of Rouen in 1492. This Morin was by far the most important of the Norman stationers and printers, and he appears to have excelled in the printing of service books, for he was employed by printers and publishers from all parts to print the service books for the special uses of the towns where they resided.

For England he pr... d altogether six service books in the fifteenth century. Three *Missals*, two *Breviaries*, and a *Liber Festivalis*, and of these the *Missal* of 1492 is the earliest. The two copies known of this book, both slightly imperfect, are in the British Museum and the Bodleian. It contains, like the earlier edition printed for Caxton, two full-page engravings before the Canon of the Mass, not one only, as is more generally the case.

The two later *Missals* which he issued, one without date but about 1495 and another dated 1497, appear to have been mixed up by all writers. The undated edition appears the rarest, for the only copy which I have noted is in the British Museum. Of the dated edition I have notes of five; one at Windsor in the Royal library, one in St Catharine's College, one at Chatsworth, one in the Aberdeen University Library, and the fifth at Kinnaird Castle. I owe my knowledge of the existence of this last copy to almost the last book in which one would seek for bibliographical information, that handy work of reference *Who's Who*. Both editions are very handsome books, remarkable for their fine titles and initial letters.

Of the two *Breviaries* which Morin printed the earliest is dated 1496, and the only copy known is in the University Library at Edinburgh, to which it was bequeathed in 1577 by Clement Litill, who left a number of valuable books to that library, of which he was practically the founder. It is a magnificent folio volume of 437 leaves, and contains a fairly full imprint, which after a deal of very grandiloquent language tells us that the book was printed at the cost of Jean Richard, "by the industry of that man skilled in printing, Mr Martin Morin, a not unworthy citizen of that great city Rouen." Morin's colophons I may note rarely err on the side of modesty. The Jean Richard mentioned was a stationer of Rouen, and one who appears to have had considerable dealings with England. I do not think he was a printer, as is often stated, and he describes himself as a dealer in books, not a printer, using sometimes the word merchant of books and sometimes the word stationer.

It was for him that Morin printed in 1499 an edition of Mirk's *Liber Festivalis* and *Quattuor Sermones*, a copy of which is in the Sandars collection in the University Library. For him also, in 1500, a Sarum *Manual* was printed by Petrus Olivier and Joannes de Lorraine, of which there is a copy in the Bodleian, and during the early years of the sixteenth century a considerable number of service books for the English market were printed at his expense.

The names of a number of early stationers who probably traded between Rouen and England are to be found in the imprints of the early Sarum *Missals*, for as the printing of them entailed a good deal of expense a number of booksellers would combine to pay for the edition. Rouen seems to have been, amongst all the towns of France, the most connected with England as regards the book trade. It was there many of our printers, as well as the first Scottish printers, learned their art or obtained their materials, while stationers from that town crossed over and sold their books in this country.

We know that Ingelbert Haghe, the publisher of the Hereford *Breviary* of 1505, came over himself and sold books at Hereford and in the country round. On the fly-leaf of a Bible formerly in the library of Gloucester Cathedral is a Latin inscription which runs:

JAMES RAVYNELL

"I gave to the Hereford bookseller called Ingelbert for this and the six other volumes of the Bible 43 shillings and fourpence, which I bought at Ludlow the year of our Lord's incarnation 1510, about the day of the Lichfield fair." Whether the Bible is still in the Gloucester Cathedral library I do not know, but the fly-leaves which once belonged to it are in a bundle of scraps in the Bodleian.

Another Rouen printer issued in 1495 an edition of the *Liber Festivalis*. His name was James Ravynell, and this is the only book that he is known to have printed. It is an exact copy of the edition printed by W. de Worde in 1493 and '94, and the type used in it has a very clean and new appearance. At the end is a device with the initials P. R., which looks as though it might have been made for another member of the family, though we know of no other printer of the name. The fact that he uses the English form of the Christian name in the imprint, "By me, James Ravynell," looks as if he was an Englishman who had migrated to Rouen.

The device consists of the initials P. R. on a shield suspended by a belt from a tree and supported by two muzzled bears. Below the shield two birds hold up a wreath. Round the whole runs the text: "Junior fui etenim senui et non vidi justum derelictum nec semen ejus querens panem." The name Ravynell is a curious one, and may be a corrupted spelling of a commoner name, though it is still borne by some families of Huguenot descent.

Another mysterious book, which from its type may very well have been printed at Rouen, is an edition of the little grammar called *Parvula*. It consists only of four leaves, and the only copy known is at Manchester. The book ends: " Here endeth a treatise called parvula, for the instruction of children. Emprentyd by me Nicole Marcant." In the exasperating way common to some printers both the date and place of printing are omitted. As to the date I am inclined to put it before 1500, but the place is more difficult to settle. Nicole Marcant is an unknown printer, but may very well be a member of one of the numerous families of Marchand or Mercator, for there were several printers of that name, though none so far as I know named Nicholas.

If we except the *Missal* printed for Caxton in 1487 it was not until 1494 that the Paris printers began to work for the English market, and the books they produced were almost all liturgical. The only exceptions are three editions of grammatical works, two of the *Liber Equivocorum* wrongly ascribed to Joannes de Garlandia, and one of the *Liber Synonymorum*, the first two printed by Baligault and the last by Hopyl.

The first liturgical book was an edition of the Sarum *Breviary*, printed in 1494 by Pierre Levet. For a long time only one copy was known, that in the library at Trinity College, Dublin, but not long ago the University Library was fortunate enough to secure a second example, a very beautiful copy in its old binding.

In one thing the Paris printers excelled all others, and that was in the production of books of Hours. These were turned out in the last few years of the century by hundreds of thousands, and though they are now of very common occurrence and very often of little interest, they are still much sought after by certain classes of collectors, especially those who like what they call pretty books. Of course, when these books were printed for the use of out of the way places they have often great liturgical interest, and being printed no doubt in small quantities are very rare. The English service books having been relentlessly destroyed at the Reformation are very rare indeed. Altogether in the fifteenth century twenty-seven editions of the Sarum *Horae* were printed, fifteen in England, one at Antwerp, one at Venice, and ten in Paris. Nine English editions were printed before one was issued at Paris, but these latter when once they got a footing in England easily defied competition. The changes in the text of these books during the last ten years of the century are very curious and interesting. The *Horae* was not a service book proper, but a manual of private devotion, and so long as it contained certain fixed and definite parts additional prayers could be added at will. Consequently the editions vary greatly, and each publisher seems to have aimed at inserting new and popular prayers, and by 1500 the book had increased to almost double the bulk of its forerunner of ten years earlier.

In speaking of these books there is one point on which a word of warning may be said. And that is about dating editions which have no date in the imprint. All such are usually put down to 1488, which is the first date printed in the calendar of moveable feasts. As this calendar was made out for a nineteen year cycle running on to 1508 it was naturally not reprinted for many years, and therefore is no test for dating the printing of the book. The ten editions printed at Paris are the work of about five printers, of whom the most important was Felix Baligault.

The study of these French books of Hours is not an easy one, as there is so much confusion between printers and publishers. In some cases I am afraid the publishers used the words "printed by" in a quite unwarrantable manner, and claimed to have produced books which they had done nothing more than pay for. Then again quite half the editions produced for sale in England are without any imprint, so that we are left to conjecture who was the printer from the type or cuts used in the books. To further bewilder us, sets of cuts passed from printer to printer, and are very untrustworthy guides in assigning books. If one printer issued a *Horae* with a fine set of cuts they were promptly copied by his rivals, who in their turn sold their old sets to less wealthy printers, in fact some sets of cuts change hands almost every year.

These books are all got up in the same style, the text surrounded on every page by deep borders containing figures of saints and martyrs or pictures from the Dance of Death.

One unique edition, printed by Jean Poitevin about 1498, was picked up lately in Ireland and bought by the librarian of Trinity College, Dublin, for a small sum.

A service book of great interest is the first edition of the Sarum *Manual*, of which the only known copy is in the library at Caius College. It bears on its first leaf a Latin inscription stating that it was given to the College of the Annunciation of the Blessed Mary at Cambridge by Humphrey de la Poole, son of the Duke of Suffolk, for the use of the college, in September,

1498. The book is a folio of 164 leaves, beautifully printed in red and black by Berthold Rembolt of Paris. It has no date, but the Greek in the printer's device reads ΧΕΡΕΘΗΚΙ, and must therefore be after 1496, when it read ΧΕΡΕΘΙΚΗ, and as the book was presented in 1498 we may fairly safely fix the date of printing about the beginning of 1498. Unfortunately the last leaf is missing, which may have contained an imprint giving the exact date and stating for whom the book was printed.

The last service book to be noticed is a Sarum *Missal* printed by Jean du Pré at Paris in 1500. Unfortunately all the copies of this book are imperfect, though from the three copies known an exact collation can be made. Another *Missal* was printed at Paris in the same year by Higman and Hopyl for two unknown persons, I.B. and G.H.

All these service books though most interesting liturgically are almost the most uninteresting class of book to the bibliographer. They were issued by well-known printers, and are hardly different from the great mass of foreign service books. From them early in the sixteenth century, however, we derive a good deal of information about the stationers, especially as regards the provincial presses; for in the case of a town like York hardly anything seems to have been printed beyond liturgical books.

So far the books we have been speaking of have been for the most part in Latin, with some sentences here and there in English, printed, of course, for the English market, but not of much interest from the point of view of literature. But we now come to another small group of English books, printed entirely in English, of very much greater interest.

In 1492 and 1493, when, just after the death of Caxton, the English press was almost at a standstill, Gerard Leeu of Antwerp printed four English books of considerable interest. Three of them were reprints of books already printed by Caxton, Lefevre's *History of Jason*, the *History of Paris and Vienne*, and the *Chronicles of England*. The fourth book was *The Dialogue or Communyng between the Wise King Solomon and Marcolphus*. Of this there does not seem to have been any other English edition, though many Latin ones were printed in the fifteenth century, and it is possible, though hardly probable, that Caxton might have printed an edition which has entirely disappeared.

Lefevre's *History of Jason* is a small folio of ninety-eight leaves, illustrated with a number of half-page cuts clearly made to illustrate the book in which they first appear. They were used in several editions of the *Jason* in different languages, the earliest in Dutch having been printed by Bellaert at Haarlem about 1485. There are copies of the English edition at Trinity College, Dublin, and in the library at Chatsworth, and a third copy, slightly imperfect, is in the University Library.

The *History of Paris and Vienne*, which was printed exactly three weeks after the *History of Jason*, is a still rarer book, only one copy being known, which is in the library at Trinity College, Dublin. It, like the *Jason*, is illustrated with a series of half-page wood engravings, which Mr Conway, in his *History of the Woodcutters of the Netherlands*, conjectures to have been originally used in an edition printed by Bellaert at Haarlem, which has now entirely disappeared, and then to have passed from his possession into the hands of Gerard Leeu. It is a small folio of forty leaves, and the copy at Dublin is bound up with the *Jason* and the *Chronicles*.

The next book to be noticed, the *Dialogue or Communyng between the Wise King Solomon and Marcolphus*, is very interesting, being the only English edition of this version of a widespread and popular story. It tells how Solomon, seated on his throne, is confronted by Marcolphus, a misshapen rustic who answers with a certain coarse wit the questions put to him by the king. Later on the king visits Marcolphus, who in his turn comes to reside at court, but his behaviour there is so insolent that the king can hardly put up with it. After a series of escapades Marcolphus is banished from the court, and finally sentenced to be hanged. He is allowed as a favour to choose his own tree, and consequently he wanders with his guards through the Vale of Josaphath to Jericho, over Jordan, through Arabia and the wilderness to the Red Sea, but "never more could Marcolf find a tree that he wold choose to hang on." The curious result of this is that he went home and lived happily ever afterwards.

The book itself has only one illustration, which is used twice, on the recto and verso of the title-page, representing Marcolphus and his wife Polycana standing before Solomon, who is seated upon his throne. This cut found its way over to England, and was used by several successive printers for editions of *Howleglas*.

The only copy known of *Solomon and Marcolphus*

is in the Bodleian, and was in a volume of tracts bequeathed with his library by Thomas Tanner, Bishop of St Asaph. The volume contained originally five separate pieces. Two by Wynkyn de Worde, the *Three Kings of Coleyne* and the *Meditations of St Bernard*, two by Caxton, the *Governayle of Health* and the *Ars moriendi*, and the *Solomon and Marcolphus*. I am sorry to say that the two Caxtons have been cut out of the volume and bound separately.

The last of the four books to be noticed is the edition of the *Chronicles of England*. While the *Chronicles* were being printed Gerard Leeu died, or perhaps it would be more correct to say was murdered. One of his workmen named Henric van Symmen, who was also a type engraver, struck work and determined to set up in business on his own account. This led to a quarrel, and blows succeeded words. The workman, it appears, in the course of the quarrel struck Leeu a blow on the head, and this proved so serious that he lay very ill for three days and then died. The workman was promptly secured and brought up for trial for the killing of his master, but it was probably considered that he had received a certain amount of provocation, and his punishment took the form of a fine. He was sentenced to pay into the Duke of Burgundy's exchequer the sum of forty guelden. Gerard Leeu seems to have been a good master and a kindly man if we may judge from the colophon put to the *Chronicles*: " Enprentyd In the Duchye of Braband in the towne of Andewarpe In the yere of our lord M.cccc.xciii. By maister Gerard de Leew a man of grete wysedom in all maner of kunnyng: whych nowe is come from lyfe unto the deth, which is grete harme for many a poure man. On whos sowle god almyghty for hys hygh grace haue mercy. Amen."

The book contains no illustrations beyond a woodcut of the arms of England on the title-page.

Leeu seems to have intended to print more English books, for the type in which all but the *Chronicles* are printed was a special fount cut in imitation of English type, with a curious lower-case d for use when that letter occurred at the end of a word. His death, so soon after the cutting of the type, put an end to all such plans. The custom, however, of printing English books at Antwerp revived at the very beginning of the sixteenth century, for Adrian van Berghen printed an edition of Holt's *Lac Puerorum*, and John of Doesborch issued a whole series of English popular books, some of them remarkably curious.

Among the stationers who came to England from abroad the most important was certainly Frederick Egmont. He was probably a Frenchman, but his printing was mainly done in Venice, and he seems to have been the agent of the Venetian printer Johannes Hertzog de Landoia. From this Venetian press came a large number of service books for English use, editions of the *Breviary* and *Missal*. The Sarum *Horae* on the other hand is only represented by one edition, issued about 1494, of which only a few leaves are known.

Egmont during his earliest years as a stationer was connected with no press except that of Hertzog, and we do not know of any books by this printer produced for any other English stationer, so that as regards liturgical books for English use known to us only from fragments we are justified, I think, in attributing to Egmont as stationer such as we can determine from their type to have been printed by Hertzog.

The first book in which his name occurs is an edition of the *Breviary* according to the use of York, of which the only known copy is in the Bodleian, having been originally in the great liturgical collection of Richard Gough. It is a small thick octavo of 462 leaves, and was issued in May, 1493. Two if not three editions of the Sarum *Breviary* in octavo were printed about this same time, but we know of their existence only from fragments discovered in bindings. Fragments of one edition are in a binding in the library of St John's College, Cambridge, of another in a binding at Lambeth, while some leaves of probably a third edition are in the library of Corpus Christi College, Oxford.

In 1494 Egmont had commissioned Hertzog to print for him two editions of the Sarum *Missal*, one in folio, the other in octavo. The folio edition is of great rarity, but there is a beautiful though slightly imperfect copy in the Sandars collection in the University Library. The title-page is wanting and also the leaf containing the engraving of the Crucifixion which should precede the Canon of the Mass. In the imprint we are told that the book was finished on the 1st of September, 1494, by John Hertzog de Landoia for Fredericus de Egmont and Gerardus Barrevelt. This Gerardus Barrevelt was clearly a partner of Egmont's as their initials occur together in the device on the title-page. This device is remarkable for the delicacy of its execution. It consists of a circle divided by a perpendicular line produced beyond the top of the circle, the projection being crossed by two bars. In the left-hand half of the circle are the initials and mark of Egmont, in the right those of Barrevelt. The whole is enclosed in a square frame, and the background contains sprays of leaves. It so resembles in style and appearance the mark used by the printer John Hertzog that we may be pretty certain it was cut under his supervision at Venice.

The octavo *Missal* of 1494, a much commoner book than the last, was issued in December. On the last leaf is Hertzog's mark and the words, "Fredericus egmont me fieri fecit." There is no mention of Barrevelt, and the double device does not occur in the book, which makes it appear as though this edition was printed for Egmont alone. Both these editions of the *Missal* contain exquisitely designed woodcut initials, the most graceful to be found in any early book.

In the Bodleian there is a copy of the "Pars estivalis" of the Sarum *Breviary* printed at Venice in 1495, which contains again the device of Egmont and Barrevelt, though the imprint mentions Egmont's name only. After 1495 we hear nothing more of Egmont until 1499, when he seems to have got rid of his former partner Barrevelt and joined with a man named Peter post pascha, and these two commissioned Pynson to print them an edition of the *Promptorium Puerorum*. After 1499 Egmont disappeared for a long time; we know of him working as a bookbinder, and it is probable that he stayed on for some time in England, for he is mentioned as a witness in a law-suit in London in 1502. When he does reappear it is in Paris, where he had some books printed for him about 1517–1520.

It is very disappointing that we have practically

no information about Frederick Egmont, for it is clear from the number of books that he had printed for him in Venice that he must have been a stationer of very considerable importance. The colophons of his books give, beyond his mere name, no information whatever about him: we do not even know in what part of London or under what sign he lived. The stationers seem always to have settled in St Paul's Churchyard, and I cannot help thinking that part of that district may have been "in the liberties," as it was called, of some church. Though the Act of Richard allowed foreigners to come over and trade, yet I do not suppose his Act could override the rights of the trade guilds. It certainly did not in York, for there a stationer must be a freeman by right or by purchase before he could carry on certain businesses, that of a stationer amongst the number, within the city. There were, however, certain liberties where an alien could live and trade; and we find at York that their earliest stationer, Gerard Wanseford, does not appear in the city register. Having taken up his abode within the liberty of St Peter, he was privileged to carry on business there without being a freeman of the city.

In the same way in London, I suppose, the various trades had their rights and could prevent foreigners from competing, except they resided within the liberties. Of course there was a Stationers' Company in London in the fifteenth century, though unfortunately most of the records relating to it have disappeared, and it would protect its own members. We see in the early bindings how ostentatiously the binders who were freemen decorated their bindings with the arms of London, and there is no doubt that as far as trading in the City was concerned the foreigner was considerably handicapped in comparison with the freeman.

We know from the few early documents remaining that the London Company of Stationers was a powerful and important body, and the members of it must certainly have enjoyed certain privileges.

Nicholas Lecomte was another stationer who appears to have been settled in England by 1494, in which year, so far as I know, his first dated book appears. M. Madden, a French writer on early printing, in the fifth volume of his *Lettres d'une Bibliographe*, speaks of Hopyl having printed a book for Lecomte in 1493. Several times in writing to him I asked for some information about this book, its whereabouts or its name even, but though he sent always voluminous replies to my letters, he never would touch on this particular point. I think, therefore, we may consider that this 1493 book never existed, and take the 1494 book as the first. This was an edition of the *Liber Synonymorum*, printed by Hopyl, of which there are copies in the University Library, the British Museum, and the Bodleian.

In the imprint Lecomte is described as living in London by St Paul's Churchyard at the sign of St Nicholas. His device depicts St Nicholas restoring to life the three children who had been killed and pickled, a favourite subject of the early bookbinders.

I think it is worth noting here, that so far as I can discover the sign of a house was not in any way permanent, but could apparently be changed at will. I noticed this in reading through a catalogue or *précis* of some thousands of deeds relating to property in London at this time and a little earlier. We find endless notices of houses with changed signs, "the tenement now called the Rose, formerly the Lion," the "house called the Bull, formerly called the Rose," and so on. Naturally if a house got celebrated for any reason it would be politic to keep the sign, but there seems to have been no compulsion to do so.

In 1495 an edition of Mirk's *Liber Festivalis* and *Quattuor Sermones* was printed by Hopyl for Lecomte. This contains Lecomte's device at the end of the *Liber Festivalis* and a curious device at the end of the *Quattuor Sermones*, used sometimes by Hopyl, but which does not bear on its face any appearance of having been made for him.

At the time when this book was printed Hopyl had in his office as press corrector an Edinburgh man called David Lauxius, the earliest Scotchman we know of employed in a printing-office. He afterwards became a schoolmaster at Arras, and appears to have been a man of considerable ability, and a friend of the celebrated Parisian printer and editor, Badius Ascensius, who addresses to him some of the prefatory letters in his grammars. What Scotch name is represented by the Latin Lauxius no one has yet been able to determine.

The last book printed for Lecomte was printed at Paris by Jean Johannot, and is an edition of the Sarum *Horae*. It is a book of very great rarity, but there are two copies in Cambridge, one in Trinity College, and the other in the Sandars collection in the University Library, the latter containing a small supplement not found in the other copies, and which was not originally intended to form part of the book, since the prayers in it are not referred to in the list of contents. The imprint is curious; it states that the edition has been revised and corrected in the celebrated University of Paris, and printed for Nicolas Lecomte of that University, settled for the time being in England as a merchant of books. I do not know whether this means merely that he was educated at the University or whether he was one of the privileged stationers attached to it, though in the latter case he would hardly have come to settle in England. Like Frederick Egmont, Lecomte was also a bookbinder.

Before the end of the century another stationer was settled in England whose name we know, John Boudins. We know of only one book printed for him, an edition of the *Expositio Hymnorum et Sequentiarum* of Salisbury use, which was printed at the beginning of 1502 by Bocard of Paris. Boudins was probably then an old man, for his will is dated the 11th of October, 1501, and it was proved on the 30th of March, 1503. He lived in the parish of St Clement's, Eastcheap, and was apparently a naturalised Fleming, and an immigrant from Antwerp.

A great difficulty in the way of tracing these stationers, especially those from the Low Countries, is the very sparing use they made of their proper surnames. In legal documents such as wills or letters of denization the formal name would be given, whereas in ordinary parlance and in the imprints of books they would be spoken of by a kind of nickname taken from the town from which they came, like William de Machlinia, Wynkyn de Worde, and so on. So that we should probably find, if we had more information on the subject, that in many cases two men who are treated as different may turn out to be only one man under two names. The number of stationers that existed at this time in England was probably very large, and it is sad to think that our information on the subject is so meagre. Of course unless the stationer was wealthy enough or in a good way of business he would not be able to commission whole editions of books from a foreign printer, and therefore he would not have his name in the imprint. Then again the greater part of a stationer's stock would consist of foreign books which were not necessarily printed for England. For information of this class we can only look to manuscript sources, accounts kept by the bookseller, lists of imported books, and so on.

There exists, for instance, a list of books for sale at Oxford in 1483 by Thomas Hunte, which has been edited by Mr Madan for the Oxford Historical Society. At the head of the list is the following sentence in Latin: "Here follows the inventory of the books which I, Thomas Hunte, stationer of the University of Oxford, have received from Master Peter Actors and John of Aix-la-Chapelle to sell, with the price of each book, and I promise faithfully to return the books or the money according to the price written below as it appears in the following list." The two men mentioned were travelling stationers from London, supplying so much stock to the bookseller on a system of sale or return.

A document such as the *Day-book of John Dorne*, the journal or account-book of an Oxford bookseller in 1520, which was edited by Mr Madan for the Oxford

Historical Society, and about which Henry Bradshaw wrote his *Half-century of Notes*, the last piece of work which he finished, is a find of the utmost importance in our subject, and it is perhaps not too much to expect that more documents of this kind may be forthcoming. In the account-book we notice that after the 21st of May up to the 3rd of August there is an entire blank, and Dorne begins his account-book again "post recessum meum de ultra mare." I think we should be safe in concluding that these months were spent abroad on business and in the purchase of books.

Sometimes such information is found amongst the waste leaves used to make boards for bindings. The University Librarian read a note before the Antiquarian Society here giving an account of a letter on business matters written from a foreign printer to John Siberch, the first printer in Cambridge, which was found among other waste matter used to make the boards of a binding now in Westminster Abbey Library, and letters of bookbinders have been found in the same way.

We have not, unfortunately, any book however meagre on this subject which might serve as a basis on which to build up information. Isolated facts turn up occasionally here or there, but there being no regular place for us to put them they drop out of sight again. And it is only when we have collected a number of these facts and begin to find the links that piece them together that we can arrive at any definite knowledge of the subject.

I do not suppose we may expect to find much new information from books themselves, but from manuscript sources a good deal may yet be discovered. Within the last year or two many documents relating to stationers and printers of the early sixteenth century have been found at the Record Office, and there must be many more still to be found there; besides, the documents in the Record Office are only a part of our great collections.

However, as I said before, what we most want is an account as full as possible of the booksellers and stationers up to 1535, giving us all the information that has yet been discovered, to serve as a groundwork for what may be found in the future.

LECTURE IV.

THE BOOKBINDERS.

FROM the very earliest times the bookbindings produced in England were remarkable for their beauty and richness. The finest were of gold, ornamented with gems, but their value has led to their destruction, and I do not think that there is any early binding of this class now in existence. Leather was very soon recognised as a suitable material for book-covers, being easily worked and capable of receiving a considerable amount of ornament. The earliest leather binding known is on a beautiful little manuscript of St John's Gospel, taken from the tomb of St Cuthbert, and now preserved in the library of Stonyhurst College. It is of red leather, and the centre of the side is ornamented with a raised ornament of Celtic design, while above and below are small panels filled with interlaced lines, executed apparently with a pointed tool and coloured yellow. This binding is generally considered to be of the tenth century, though there are some reasons for thinking that it may have been executed later, but if this is so the present binding must have been copied from an earlier one.

Excellent as the early work had been, that of the twelfth and thirteenth centuries is unsurpassed. The leather bindings executed at Durham for Bishop Pudsey between 1153 and 1195 are marvellous both for their detail and for their general effect. It was the custom of binders of this period to build up a bold and effective pattern covering the whole side of the book by means of a large number of dies, beautifully engraved with different designs. On the four volumes of Bishop Pudsey's Bible, now in the Cathedral library at Durham, no less than fifty-one different dies are used, and when we remember that Bishop Pudsey was one of the great builders of the cathedral, it is not surprising that the ornamentation on the dies used in these bindings should resemble the carved work in the cathedral. There are in the Cathedral library seven of these early bindings, and, unfortunately, they have suffered a considerable amount of mutilation at a not very remote date, for visitors on payment of a small gratuity to the person who looked after the library were allowed to cut out with a penknife one of the stamps to keep as a curiosity. A few more Durham bindings, easily recognised by the dies, are scattered in different libraries in London and in France.

At Winchester, also, and London very beautiful work of the same class was produced, the circular form of decoration being very much made use of. Perhaps the finest piece of Winchester work now in existence is the binding of the *Winchester Domesday Book* in the library of the Society of Antiquaries, of which a facsimile was published in the illustrated catalogue of the exhibition of bookbindings at the Burlington Fine

Arts Club. Some very fine work, too, probably executed at Winchester, is to be found on some manuscripts in the library of the Faculty of Medicine at Montpellier executed before 1146 for Henry, son of Louis VII of France.

The metal dies with which these bindings were stamped were practically indestructible, but it is curious to notice that they hardly ever appear to have been used after the twelfth and early thirteenth centuries. In Westminster Abbey library is a copy of the *Epistolae* of Ficinus, printed in 1495, which has its covers ornamented with early Winchester stamps, and another binding worked with twelfth century dies is on a fifteenth century printed book in the library of Corpus Christi College, Cambridge.

In all these early bindings one is especially struck with the extraordinary taste and balance in the decoration. The dies themselves are beautiful, and the pattern in which they are built up is also beautiful, and yet neither are unduly emphasised. In later bindings the die became smaller and less finely cut. It was not intended to be decorative in itself, but only to help to build up patterns, and the bindings in consequence lose much of their interest.

Oxford, I believe, is generally credited with clinging somewhat strongly to old traditions, and certainly its bookbinders did so in the fifteenth century. From the earliest times bookbinding had been considerably practised there and continued without a break, and no doubt that is why the old styles lingered for so long. The bindings produced there towards the end of the century form the connecting link between the old styles and the new. They represent the last survival of the early English school of work, that very distinctive English style which depended so much on the disposal of dies into large circles, or parallelograms one inside the other, such as we find in the Winchester and Durham bindings of the twelfth or thirteenth centuries. That this circular work was not the haphazard freak of a single binder we can see from the fact that several of the dies are wider at the top than at the bottom, so that when placed together side by side they would naturally work round to a circular form, like the stones forming the arch of a bridge. These dies are in many cases foreign in design and may have been introduced by Rood, the first printer, but the style of binding is essentially English. Some bindings of a rather similar appearance, though never with any circular ornament, were produced in the Low Countries. On nearly all Oxford bindings will be found little groups of three small circles, so small that they might have been done with the end of a watch-key, and arranged in a triangle. This ornament I have never seen on any but Oxford work. One habit connects the Oxford binders with those of the Low Countries, and that is their habit of always, when possible, lining the boards of the binding with leaves of vellum rather than paper. All the other English binders used paper generally for this purpose. It is owing to this custom of using vellum that many copies of *Indulgences* issued by the early printers have been preserved, for, as they were only printed on one side, the binder could paste them down with the printed side next the boards and the clean side outwards. An Oxford binding with an inscription stating that it was bound in "Catte Strete" in 1467 was formerly in the British Museum: the manuscript which it covered has been rebound and the old binding has disappeared.

Caxton, as one would naturally expect, followed the style of binding which he had become used to during his residence at Bruges, though it is interesting to notice that one at least of his dies was directly copied from early London work and applied in the same manner. His general method of covering the side of his binding was to make a large centre panel contained by a framework of dies or lines running about an inch from the edge of the side and intersecting each other at the corners as in the frame known as an Oxford frame. The large panel thus produced on the side was divided into lozenge-shaped compartments by diagonal lines running both ways from the frame, and in each of these compartments a die was stamped. The die most commonly found on his bindings is a square one with some fabulous winged monster engraved upon it, and this very die we find later in the hands of a stationer in London named Jacobi. The broad frame was often made up by repetitions of a triangular stamp, pointing alternately right and left, and containing the figure of a dragon. This stamp is interesting, not only because the use of a triangular stamp was very uncommon, but because it was an exact copy of one used by a London binder about the end of the twelfth century. Very few of Caxton's own books in their original binding have come down to our time, but there is a copy of the second edition of the *Liber Festivalis* in the British Museum which was clearly bound by him, and the

Boethius which was found in the Grammar School at St Alban's was also in its original cover. The *Royal Book* formerly in the Bedfordshire General Library, and now in the collection of Mr J. Pierpont Morgan, is in an absolutely similar binding to the *Liber Festivalis* in the British Museum, ornamented with the same die, while the boards were lined with two waste copies of an *Indulgence*. Caxton's bindings were invariably of leather; he never used vellum as many writers have stated. Blades, who was amongst the number, refers to a vellum-bound Caxton in the Bodleian, and states that it is the original binding; but had he examined the book more carefully he would have found that it was made up from two copies, and that the binding therefore could not well be original. Indeed the particular binding was put on in the seventeenth century while the book belonged to Selden. Selden's bindings had good need to be flexible, for one of his customs did not tend to improve bindings. He used to buy his spectacles, like the youth in the *Vicar of Wakefield*, by the gross, and whenever he stopped reading a book he put in the pair he happened to be using to mark the place. It was quite a common thing, soon after his library came to the Bodleian, for spectacles to drop out of the books as they were taken incautiously from the shelves.

Of course the number of bindings which can with certainty be ascribed to Caxton is necessarily small, we can in the first place take only those on books printed by him, and which contain distinct evidence from the fragments used in the binding that they came from his workshop. By means of the stamps used on these we can identify others which have no other materials for identification. Caxton used sometimes wooden boards in his bindings and sometimes waste leaves of printed matter pasted together. These pads of old printing frequently yield most valuable prizes. The copy of Caxton's *Boethius*, found by Blades in the library of the St Alban's Grammar-school, had its boards made of printed matter, which, when carefully taken to pieces, were found to be made of fifty-six half-sheets of paper, forming portions of thirteen books printed by Caxton, three of which were quite unknown.

Caxton's binding stamps passed with his printing material to his successor, Wynkyn de Worde. I found in a college library at Oxford a book with these stamps, evidently bound by De Worde, and the boards were lined with waste leaves of three books printed by him, one being unknown, and one by Caxton. De Worde's bindings are the least easily identified of any in the fifteenth century, for beyond these few dies of Caxton's there are none that can definitely be ascribed to him, and even the various bindings that might be ascribed to him from the fragments found in them vary so much in style and decoration that it seems impossible that they could have all come from one shop. Perhaps he had really no binding establishment of his own, but got such work as he required done by others.

Wynkyn de Worde, as we learn from his will, employed several binders. He left bequests to Alard, bookbinder, his servant, and to Nowel, the bookbinder in Shoe Lane. James Gaver, who was one of his executors, was one of the large family of Gavere, binders in the Low Countries, and though, when he took out letters of denization on his own account in 1535 he is described as a stationer, no doubt he was also a bookbinder. The square stamp with a dragon, which had belonged to Caxton and which must have passed to De Worde, found its way early in the sixteenth century with other dies of Caxton's into the hands of another stationer, Henry Jacobi.

The bindings which were produced by Lettou and Machlinia, so far as we are able to identify them, are very plain. The sides are divided by diagonal lines into diamond-shaped compartments, and in each is stamped a small and uninteresting die. The Latin Bible in the library of Jesus College, Cambridge, which has every quire lined with slips of vellum, portions of two cut-up copies of Lettou's *Indulgence*, and presumably bound by him, has its binding ornamented with diagonal lines within a frame formed of square dies containing the figure of a fabulous animal. In the diamond-shaped compartments formed by the diagonal lines is a small impressed cinquefoil. Another Lettou binding, on the copy of the *Wallensis* printed by him in 1481 in the Bodleian, is ornamented simply with diagonal lines, but has no small stamps.

There is another English binder of this time whose name we do not know, who produced some very good work. Bradshaw, I think, considered that he worked at Norwich. There are a number of his books in Cambridge libraries, and he used very often a red-coloured leather, which is common in Cambridge bindings. His dies are Low Country in type, and very much resemble those used at Oxford, but his work can be recognised by two peculiarities. He always ruled two perpendicular lines down the backs of his books, and always ornamented the ends of the bands, the bands being those ridges on the back where the leather covers the string or cord on which the quires are stitched. Where these bands ended on the sides he printed a kind of ornament of leaves. He also, like the Oxford binders, almost always lined his boards with vellum. His dies, about eighteen in number, are well engraved, one in especial representing two cocks fighting, being very finely executed.

Pynson's earliest bindings are as a rule very plain. Like the other binders of the time he ruled diagonal lines across the sides of his books, and put a small die in each division. Sometimes he did not even use a die, but contented himself with plain lines, as, for instance, on the copy of his first dated book of 1492 in the British Museum. His bindings, like Machlinia's, are very plain, and the dies used are small and poor.

Another binder, perhaps at St Alban's, produced bindings not unlike Pynson's, but he is identified by a small circular die which he used, which has on it the figure of a bird.

Another binder whose initials were W. G., but whose name we do not know, produced a large number of bindings in the fifteenth century. It is from his bindings that all the fragments of Machlinia's *Horae ad usum Sarum* have been recovered, for he seems to have used up a copy for lining his boards, and luckily several books bound about that time have been preserved. Bradshaw found a curious case of the preservation of two volumes bound in the same workshop about the same time. In the library of St John's College, Cambridge, is a copy of a book printed at Nuremberg in 1505, which has in its cover some leaves of early Oxford printing. In the library at Corpus Christi College, Cambridge, is an exactly similar binding on a book printed at Lyons in 1511, which also contains some early Oxford leaves. Now it is clear that the same man must have bound these books about the same time, because we find in both, along with the refuse Oxford leaves, some leaves from one and the same vellum manuscript.

There is one English binder, who worked before the end of the fifteenth century, who is distinctly worthy of special mention on account of the striking originality of his method of decoration and designs. His name, unfortunately, we do not know, but as one of his most frequently used dies represents a balance or pair of scales it has been conjectured that this may be a rebus on his name, such as many binders used, and that he was called "Scales." Two volumes executed by this binder are known, which were done for a certain William Langton, and the centre panel is ornamented with a rebus on the name Langton, the letters Lang over a barrel or tun, while the rest of the side is filled up with little stamps. This Langton may perhaps be identified with the William Langton who was a prebendary of Lincoln and afterwards of York at the end of the fifteenth century. Another even more curious binding by this same man is in the library of Corpus Christi College, Cambridge. He has disposed his dies so as to form a large heraldic shield, covering the whole side of a folio volume, a style of adornment quite unique so far as I am aware, and as an ornament extremely effective, though I am afraid the heraldry is hardly sufficiently accurate to enable us to determine for whom the volume was bound.

The bindings that I have spoken of so far were all produced in a slow and laborious manner, as each die had to be impressed separately. Towards the end of the fifteenth century, however, when the printers in England began to issue books of a small size, a new system of binding was introduced, by which the labour of the binder was very considerably lessened, while the amount of decoration applied was increased.

The invention and use of the panel stamp, that is of a large stamp which should ornament the side of the book with one picture, was a great step forward. It was a great advantage commercially as it saved much time, and in some ways it was an advance artistically. By its means the whole side of the book was ornamented at once, instead of by a series of dies impressed one after the other. And as the working out of a binding had ceased to be its main point and the beauty of the die itself was more emphasised, this invention did away with the building up of a pattern altogether, and depended entirely on the excellence in design and workmanship of the stamp. Mr Weale assigns the date 1367 to the earliest panel stamp known to him, produced by a certain Lambertus de Insula at Louvain, but this is only because the manuscript on which it occurs bears that date. Without some further evidence I should be inclined to think this date rather too early, and would not date any panel stamp before the fifteenth century.

There is no doubt that the binders of the Low Countries were the earliest to introduce this style of binding, and they produced very excellent work; and the earliest panel stamps we find in use in England are Netherlandish in execution, either used in this country by foreign workmen, who had come over and settled, or obtained by native binders from abroad. The earliest stamps were no doubt for the most part of metal, and therefore practically indestructible, and we know that they often passed out of the hands of their proper owner and were used by other binders, even though the name of their original owner was engraved upon them. As an example I may mention a book-cover in the Douce collection in the Bodleian, on which two stamps are impressed side by side. One has the name John Guilibert, the other the inscription, "Omnes sancti angeli et archangeli dei, orate pro nobis. Ioris de Gavere me ligavit in Gandavo." A still more marvellous example, and one almost certainly bound in England, is in the library of Corpus Christi College, Oxford. It has on the two covers, besides innumerable dies, no less than nine panels, two signed Woter Vanduffle, three signed Martinus de Predio, and four signed Jacobus, illuminator. The binding almost looks like a sample put out to show a specimen of every stamp and die in the establishment. The Woter Vanduffle stamp seems very early. I have in my own collection an English heraldic manuscript of about the middle of the fifteenth century or slightly earlier in its original binding impressed with the two panels of that binder.

In these earliest panels the inscription nearly always runs perpendicularly, either in the centre of the panel, cutting it in two, or at the side of the picture. One peculiarly distinctive feature of the earliest panels is the presence of four indentations, more or less deep and clearly defined at each corner. These were made most probably by the heads of the nails by which the metal plate was affixed to a block before used for stamping. These four marks never seem to occur in later panels, which, if they have any, have only two, considerably larger in size, one at the top and one at the bottom. It has long been a vexed question as to whether these stamps were made of metal or of wood, but it is probable that both materials were used, and that the majority of English stamps were of wood. As no heat was applied and the leather treated when it was damp and soft, a wooden stamp would be sufficiently strong, and I have found by experiment that soft leather takes an excellent impression from a wooden block. I have, however, in my own collection a binding struck from a broken plate, and the appearance of the break shows clearly that the stamp must have been of metal.

The earliest definitely English panel stamp is on a loose binding in Westminster Abbey library. It has on it the arms of Edward IV, with two small supporting angels. The rest of the binding is covered with small dies, one in the shape of a heart, the other a *fleur de lys*. It is a great pity that the book which was in this binding has been lost, as it might have contained some clue to information about the binding.

Wynkyn de Worde, in spite of his enormous business, does not seem to have ever used a panel with his name or device, at least so far not one has been found, but with other printers and stationers the case is different. Pynson used two panels. One is a copy of one of his devices, having his initials on a shield with the helmet and crest above, while around all is a floral border. The other has in the centre a large Tudor rose, surrounded by intertwined branches of vine leaves and grapes. This latter panel was a popular one, and several variations of it are to be found, all of which are probably of the fifteenth century.

The only copy at present known of the Pynson panel with his mark was acquired not long ago by the British Museum. I had known of the existence of the copy for some time, as it had belonged to a Manchester bookseller who had described it to me. He had sold the book, but had no record of the purchaser, and knew nothing of him further than that he lived in London. One day while I was working in the Museum a visitor came in with this identical book and offered it for sale. The book itself was a copy of the *Abridgement of the Statutes of* 1499. Herbert, in his *Typographical Antiquities*, describes a copy of the *Imitation of Christ*, printed by Pynson, which was in a similar binding, and perhaps that may still be in existence; but I am sorry to say that the collectors at the beginning of the present century ruthlessly destroyed all old bindings, and would not have anything on their shelves except bound in morocco or russia by Roger Payne or Charles Lewis. There is not one single old leather binding in the whole of the Spencer library, though we know that many of the books when bought were in their original covers.

Frederick Egmont, the stationer about whom I spoke in my last lecture, had several panels. The first has as its central ornament the Tudor rose, and round it are vine leaves and grapes. Round the whole is an arabesque

BINDING OF FREDERICK EGMONT.

floral border containing the initials and mark of Egmont.
This design was common at the time, there being several
other panels almost identical, one of which was used by
Pynson. Another more important panel is an almost
exact copy of the device of Philippe Pigouchet, the
Paris printer. A wild man and woman, standing on
either side of a tree covered with some kind of fruit,
bear in one hand flowering boughs while with the other
they assist in supporting a shield suspended by a belt
from the branches above them. Upon the shield are
Egmont's mark and initials. The device of the wild
man and woman was for some reason very popular at
this time and for a short period afterwards. It was
used by Bumgart at Cologne, and at Edinburgh by
Walter Chepman and Thomas Davidson. It was used
by Pigouchet and other Parisian printers, and by Peter
Treveris, who printed in Southwark at the sign of the
"Wodows," and the references to it in colophons are
very numerous. This panel of Egmont's not only bears
his mark and initials, but is inscribed on the lower
margin, "Fredericus Egmondt me f[ecit]." Three copies
only of this binding are known, a very fine copy at
Caius, a poor copy at Corpus Christi College, Cambridge,
and one in my own collection. Books which are stamped
on the front with this panel generally have on the back
a plainer panel containing three rows of arabesques of
foliage surrounded by a border, having ribbons in the
upper and lower portion inscribed with the names of
the four Evangelists. This panel not infrequently
occurs alone, without Egmont's signed panel, and may
have been left by him in England when he returned to
France.

Nicholas Lecomte, the other foreign stationer settled in England in the fifteenth century, and of whom I spoke in my last lecture, also used panels with his initials and mark. He had not a pictorial panel, but a formal one of rather Low Country type. The centre of the panel is divided into two parts, each containing four spirals of foliage encircling the figures of beasts and birds, while around all is a border of grapes and vine leaves. At the bottom of the border are the initials N.C., with a mark which almost exactly corresponds with the initials and mark found in his device in the books printed for him. A very fine specimen of this binding is in the University Library.

The majority of the early panels are pictorial, and in some cases they are very elaborate and ornamental. The pair used by a binder whose initials were A. R. are especially fine. On one side is the salutation of Elisabeth with the Almighty and the Holy Ghost above, in the top corners are the Tudor emblems, the rose and portcullis. Round the whole is a diaper border with a shield in each corner, one the arms of St George, another of the City of London, a third with two cross swords, and a fourth with two cross keys. The panel on the other side has in the centre St John standing preaching to some people who sit in the foreground. On the left is St James, on the right King David, while the lower part is taken up with a picture of David and Bathsheba. This binding is very rare, and I know of only two examples, one in my own collection and one in the University Library. This binder had several other panels; there is one in the University Library with a figure of St Roche, and there are two on a book at

Aberdeen with the Baptism of Christ, and the Annunciation.

It is curious that the subjects of these panels should have been invariably religious, scenes from the Bible or pictures of saints, and that we never find subjects from popular stories. The most frequent subject of all was, I think, the Annunciation, and then single figures of saints, St Barbara being one of the most popular. The binder very often used a panel with a figure of the saint after whom he was named. Nicolas Spering, who worked at Cambridge, has a panel with the picture of St Nicholas restoring to life the three children who had been killed and pickled by the innkeeper. Another binding which is probably English, though it might be French, has on one side St Barbara with her palm branch and three-windowed tower, and on the other the Mass of St Gregory. In the border there occurs a delightful little figure of a mermaid with a comb in one hand and a looking-glass in the other.

About 1500 a particular pair of panels came into great vogue amongst the bookbinders. One had upon it the arms of England, supported by the dragon and greyhound, the other the Tudor rose supported by angels. Round the rose runs a ribbon with the motto:—

> Haec rosa virtutis de celo missa sereno
> Eternum florens regia sceptra feret.

In the top corners we generally find shields with the arms of St George and of London, while in the base below the rose or shield occur the initials and marks of the binders. This general use of the royal arms together with the use of the arms of London points, I think, to some trade guild to which these binders belonged. Foreigners, though they might still use the royal arms, do not use the City arms, putting something else in the place, sometimes the French shield, sometimes merely an unmeaning ornament. It is a very popular but erroneous opinion held by a great many people that these bindings with the royal arms were produced for the king, Henry VII or Henry VIII as the case may be. It would be just as reasonable to imagine that all the shops with the royal arms over the door were private residences of the king. Of course, the fiction is kept up in order to increase the price of the books; "from the library of Henry VIII" looks well in catalogues. Even in the sumptuous work recently issued on the historic bookbindings in the Royal Library at Windsor this mistake has been repeated.

A very large number of these bindings exist, all very similar; but unfortunately, although in many cases they bear the binder's initials and mark, we cannot discover his name; on the other hand again we know the names of many binders, but we cannot identify their work; the mere fact that the initials on a binding agree with the initials of a binder's name does not of necessity determine that the particular binding was produced by that binder; a good deal more proof is necessary.

A certain number of these bindings have been settled as the work of a certain man in another way. When the binder was a printer, or a stationer of sufficient importance to have books printed for him, then we can identify the mark on his bindings by means of the mark used in the books.

For instance, to take an early example. We have bindings by Julian Notary, the printer, which bear his initials and mark, and the mark, of course, is the same as the one he uses in his books, while in them his name is in full. So again the work of Henry Jacobi, an important London stationer of the early sixteenth century, was traced by the mark which he uses in some of his books.

We know the names of a considerable number of early binders from the registers of the grant of letters of denization and other manuscript sources, but unfortunately we have no link between them and the bindings. In this country it was not necessary, as it was in some parts of the Low Countries, to register the design of a binding, and though many of the Low Country bindings look the same, you will find on examination that the detail varies and each design was protected.

The binder who is best known in connexion with these stamped bindings is John Reynes, whose work is by far the most commonly met with, and who is almost the only producer of stamped bindings mentioned by any early bibliographer. His best-known panel is called "Redemptoris mundi arma," and consists of all the emblems of the Passion arranged in a heraldic manner upon a shield. Reynes was certainly employed as a binder by Henry VIII, as we know from early accounts, and so far as I have seen all the copies of the king's *Assertio septem sacramentorum*, which remain in their original binding, were bound by him. It is fortunate that Reynes put his mark and name in one printed book, otherwise we should not have been able to identify him as the binder. He had also two very well-executed panels, one depicting the fight of St George and the dragon, and the other the Baptism of Christ.

The period during which these panel stamps were produced in England was roughly the forty years from 1493 to 1534. The passing of the Act against foreign workmen in the latter year had no doubt a good deal to do with the falling off of the work, but the invention of a binding tool called a roll seems to have finally put an end to the use of the panel. The roll was a tool made in the form of a wheel, which saved a very great deal of time in ornamenting the sides of a book, and which was used very widely in England during the sixteenth century. At first when the roll was broad and well cut, as the earliest examples almost always were, it produced a very satisfactory appearance, but it soon became narrower and more finely cut, and therefore showing to much less advantage on the side of a large book, and finally about the end of the century its use was almost entirely given up.

Almost the earliest and the finest of the roll bindings were those produced by the Cambridge stationers. Nicholas Spering beside his panel had several fine rolls which contain his initials and mark.

Naturally the foreign booksellers who sent books over to England found it to their advantage to put them into popular bindings, such as would attract purchasers, and many of these bindings have a distinctly English character. The exploits of St George and St Michael are favourite subjects, and are often treated in a most decorative manner. There is one specially fine example dating from about 1500, and probably Rouen work, which has St George on one side and St Michael on the other. The binder has not put his initials, but his device, which occurs on one side, is a head on a crowned shield. It is worth noticing that the material of which these foreign bindings are made is often sheepskin rather than calf, which is nearly always used in English work. One binder, whose initials were A. H. and who used the Tudor rose, though without the arms of London, produced very good work, but almost always on this sheepskin, which was not a suitable leather for giving a clear impression.

It is very interesting to watch how in the later panel bindings the lettering gradually deteriorated and became simply part of the ornament. I have three panels, all copied one from the other, and in the first the legend running round the panel is quite clear and correct. In the second the letters are confused, though the general appearance of each separate word is preserved and they can be read. In the last example letters and words are all run together, and the general result is wholly unreadable.

So, too, the old style of work with the pictures of saints or Biblical scenes was given up about 1530 for bad Renaissance patterns of pillars and classical heads, which are so uninteresting, not to say ugly, that we can hardly regret the speedy disuse of the panel stamp.

Now it must always be remembered that in England at any rate very few of these early bindings are signed, and that therefore to assign particular bindings to particular men is not often possible, but comparison may enable us to attribute them to particular districts, and even to particular places. What is wanted is that every small point about these bindings should be studied carefully and compared in different examples, because it is mainly by circumstantial evidence that we can arrive at any knowledge about them. We must class our bindings by a system similar to that lately adopted for identifying criminals. The presence or absence of one particular point merely divides a number of bindings into two divisions. This point, taken in conjunction with a second point, narrows the field immensely, and we can soon put the bindings into groups more or less accurately.

Anyone who works at all amongst old bindings will soon begin to note points which are common to certain bindings, and which most probably mean a certain thing. For instance, anyone working at the subject would soon perceive that as a rule octavo or quarto books in an English binding have three bands to the back, that is three projecting ridges on which the leaves are stitched, while foreign bindings have four or more. Of course this is not an absolute rule but it will be found correct in nine cases out of ten. To take another local instance. A very great number of early Cambridge bindings, and some that may have been produced at places not far distant, are remarkable for the curious red colour of the leather used. The binding has the appearance of having been painted over with red, and then the red almost all rubbed off again. This is probably caused by some peculiarity in the process of tanning or dressing. Whenever I see this curious red colour I promptly put down the binding as a Cambridge one, and a more careful examination generally proves it to be correct.

If the boards of the binding have a groove running down the edge you may be fairly certain that the book inside is printed in Greek. If a binding has four clasps, one at the top and bottom as well as two in their usual place, you may be sure that the binding is Italian. Most of these old bindings had clasps or ties of ribbon to keep them shut, but in nearly all cases these have disappeared, and the reason is this. In the early times books were always put on the shelves back first and with the fore edge to the front, on which was written the title of the book. Naturally, when readers wanted to take down a book they pulled at it by the clasp or ribbon till that came off, just as now-a-days when books are placed backs outwards the ordinary reader pulls them out by the top of the back, till that comes off. In many cases the ribbons were of alternate colours, a white opposite a green and a green opposite a white. Of course as soon as books were put in the shelves with the backs outwards the use of ribbons was discontinued, for it was awkward to push into its place a book with two large bows of ribbon in front. The use of ribbons you will notice has been lately revived by some faddists who have no sense of the fitness of things.

These bindings that we have been considering were of course what we should call trade bindings or publishers' bindings. Very few people seem to have had books especially bound for them, and those kind of bindings had generally gilded ornaments upon them, which are not found on early stamped work. The custom of impressing coats of arms on books did not begin until about 1535, when it was started by some Scottish collectors, the earliest known armorial stamp having been used by William Stewart, Bishop of Aberdeen.

The books specially bound for Henry VIII were ornamented in what was called the Venetian manner, that is with tools obtained perhaps from Venice, but clearly cut in imitation of those used by Aldus for his bindings; the binder of these books was the well-known stationer, Thomas Berthelet.

While these bindings and their designs afford valuable bibliographical information, the materials employed in making the bindings are also often of great importance. The boards were often made of refuse printed leaves pasted together, and were always lined, after the binding was completed, with leaves of paper or vellum, printed or manuscript. To show you how important these fragments may be, I may mention that of the books printed in England or for England in the fifteenth century no less than fifty-three are known only from fragments obtained from bindings. The great find of Caxton fragments made by Blades at St Alban's I have mentioned before. Not long ago I took to pieces the boards of a primer of Edward VI and obtained the title and some other leaves of Constable's *Epigrams*, printed by Pynson in 1520, and of which but one perfect copy is known, four leaves of a Whitinton's *Grammar* printed by W. de Worde, eight leaves of an early *Abridgement of the Statutes*, probably printed by Middleton, a perfect copy of an unknown edition of the *Ordynaunce made in the time of ye reygne of kynge Henry VI to be observed in the Kynge's Eschequier by the offycers and clerkes of the same for takyng of fees of ye kynges accomptis in the same courts,* printed by Middleton, and last an unknown broadside ballad relating to the burning of Robert Barnes in 1540, printed for Richard Bankes by that little-known printer, John Redman, who put his name only to one or at the most two known books.

From a binding in Westminster Abbey some years ago came two leaves of an unknown early Cambridge book, Lily's *De octo orationis partium constructione*, edited by Erasmus, and lately at Oxford Mr Proctor found in a binding in New College some fragments of a *Donatus Melior* on vellum, printed by Caxton, a hitherto unknown book. As Bradshaw said over twenty years ago: "It cannot be any matter of wonder that the fragments used for lining the boards of old books should have an interest for those who make a study of the methods and habits of our early printers with a view to the solution of some of many difficulties still remaining unsettled in the history of printing. I have for many years tried to draw the attention of librarians and others to the evidence which may be gleaned from a careful study of these fragments; and if done systematically and intelligently it ceases to be mere antiquarian pottering or aimless waste of time."

Of course the majority of fragments found in bindings are of no value, and should not be moved; indeed, fragments should never be taken out of bindings unless it is absolutely necessary, for by doing so the binding is almost certain to suffer some injury.

To study effectively the early English book a certain

knowledge about these early bindings is required, for the printer, as we have seen, was probably his own binder. What I said about the stationers applies also to the binders, their history is an almost unworked subject, new details are found from time to time, but we have no work on the subject to which we can add them, and our knowledge at present consists mostly of isolated facts. Bradshaw, writing twenty years ago, spoke of the subject as still in its infancy, and I am afraid that English bibliographers cannot boast of much progress. This is not, perhaps, to be much wondered at when we consider how few are willing to work on in the steady, quiet way which he practised and taught. We can do no better than follow in the path that he pointed out, add fact to fact, and detail to detail, avoiding vain theories and idle speculations, so that whatever advance we make in our knowledge of the subject, whether it be much or little, it may at any rate be accurate, and serve as a secure foundation for the work of the future.

PART II.

1501—1535.

LECTURE V.

WYNKYN DE WORDE AND THE POPULAR PRINTERS.

In the four lectures which I had the privilege of giving as Sandars Reader in the Lent term of 1899 I dealt with the printers, stationers, and bookbinders of London and Westminster in the fifteenth century. In the present series I propose to continue their history up to the year 1535.

The date which has been fixed upon is not a purely arbitrary one, but has been chosen for two reasons. In the first place, it is the year in which Wynkyn de Worde, by far the most important and prolific of all the early English printers, died, and secondly, it just includes and allows us to examine the remarkable change brought about in the English book trade by the passing of the Act relating to printers in the twenty-fifth year of Henry VIII, which came into force on Christmas-Day, 1534.

Though Westminster was joined with London in the earlier lectures, it ceased after 1500 to have any printers of its own, for in that year De Worde moved to London to be nearer the centre of trade, and he was followed almost immediately by Julian Notary. In this lecture I am going to take, besides De Worde, all the printers and stationers who were connected with him in business, and who formed what might be called the popular school of printers, who neglecting legal, political, and learned books, such as were issued by the King's Printers, confined their attention to books of a lighter and more ephemeral kind.

Wynkyn de Worde, who had probably been an assistant of Caxton's from the time of the introduction of printing into England in 1476, was certainly married and settled in Westminster by 1480, in which year his wife Elizabeth is mentioned in a deed. He inherited all Caxton's printing material and continued to lease his old house and printing-office from the Abbot of Westminster, John Esteney, in whose account-book he is entered from the year 1491 onward. Mr Scott, of the British Museum, who first made known these entries, which he had found when calendaring the Abbey muniments, noted that he was called Jan Wynkyn in them, and wrote a letter to the *Athenaeum* to point out that the printer's Christian name, hitherto unknown, was John. It appears to me that about this name there is some confusion or mistake.

Wijnand or Wynkyn is itself a Christian name, and De Worde like most other foreign printers made use of his Christian name joined to the name of the place from which he came. Thus we have Joannes Lettou (John of Lithuania), Willelmus de Machlinia (William of Malines), Jan van Doesborch, Christopher van Ruremond, John Siberch, and so on, while we have no example of a surname joined with the name of a place. In all the hundreds of colophons to De Worde's books, in his patent of denization, even in his will, there is no hint of such a name as Jan, and the combination Jan Wynkyn could only mean John the son of Wynkyn. We must presume that De Worde knew better than the Abbot what his own name was and that the Abbot's entry is the result of some confusion. I am the more anxious to point out this confusion about the name because entirely without my knowledge and after I had returned my last proof the editor of the *Dictionary of National Biography* inserted this piece of information, extracted from Mr Scott's letter, in my notice of Wynkyn de Worde, and thus made me responsible for a statement which I do not in the least believe.

De Worde continued to print in Caxton's house until some time in 1500, when he moved to the sign of the Sun in Fleet Street. The position of the house can be settled almost exactly. It is described as over against, that is, opposite the Conduit, and this was situated in Fleet Street, just where Shoe Lane entered it on the north side. De Worde's house being in St Bride's parish and near the church must have been on the south side of the street and therefore opposite the entrance of Shoe Lane. He rented two houses, one no doubt a dwelling-house, the other his printing-office, and for these he paid a rent of sixty-six shillings and eight-pence. On leaving Westminster De Worde either destroyed or parted with a considerable portion of his printing material. Some of his type used up to this time never appears again, and many of the wood blocks used to illustrate his books are found in the hands of other printers, more especially Julian Notary.

The year 1501 was apparently mostly taken up with settling into his new premises; for this year his output was the smallest during his whole career, and we know of but one dated book issued in it, a new edition of Bishop Alcock's *Mons Perfectionis* which appeared on May 27. Three copies of this book are known, in the libraries of Peterborough and Lincoln Cathedrals, and the third in the University Library, which came in the bequest of Mr Sandars.

On April 22, 1502, De Worde issued an edition of the *Manipulus Curatorum*, which is worthy of notice. It is a very small octavo, printed in a small neat black-letter, and all the copies I have seen, with one exception, have on the title-page the small square early device of the printer. The copy, however, which belonged to Richard Farmer and is now in the Bodleian, has a device of De Worde which so far as I know is found in no other book. It is, like his most common device, divided into three parts. The upper contains the sun, two planets, and thirty-six stars, the middle Caxton's mark and initials, and the lower the unicorn and Sagittarius above a ribbon containing De Worde's name in full. The engraver has made a not uncommon mistake and has engraved the large initial *C* so that it prints the wrong way about. Another device of De Worde's had the mark reversed, but that was not so obvious an error; in this case the printer seems to have thought the mistake too flaring and to have suppressed the device.

There is a curious little undated tract in the University Library which cannot be later than the beginning of 1502, describing the doings of Margerie Kempe of Lynn, a religious enthusiast who travelled about the country with an axe asking people to cut her head off. In this De Worde used, probably for the last time, the

UNRECORDED DEVICE OF W. DE WORDE.

beautiful cut of the Crucifixion which he had inherited from Caxton, but which beginning to split in 1499 had now broken in half and was in other ways more or less damaged.

In 1504 De Worde began the use of his best known device, first used in a *Grammar* of Sulpitius, of which the unique copy is in the library of Shrewsbury School. It is square in shape and divided into three parts. In the upper are the sun, two planets, and twenty stars, in the middle Caxton's initials and mark, and in the lower the printer's name on a ribbon, above which are a dog and a centaur. This device was used until 1518, when having got cracked and broken it was replaced by an almost exact facsimile. This in its turn was used until 1528, when it was replaced by a third copy. So similar to the eye are these three varieties that no writer on bibliography has noticed the differences, though they form the most valuable date test we have for De Worde's books, as it was this device he most generally used.

Towards the end of 1508 when Pynson was appointed printer to the King, De Worde appears to have received some sort of official appointment as printer to the Countess of Richmond and Derby, the King's mother. He had printed several books before this time at her request, a phrase which I suppose meant that she had helped to defray the cost, but he had not called himself specially her printer. In most of the books printed in 1509 before the 29th June, when the Countess died, he gives himself an official title, as printer to the King's mother.

Among the books printed in 1508 is one entitled

The Book of Kerving, of which there were several editions. A copy of one of these editions, which was in the collection of Rawlinson, contained the following rhyme in an old hand,

> Wynken de Worde
> Sate at the borde
> Wyth hys cosyn forde
> And kyld hym with a sworde.

Below this the learned owner has written " Whether this last writing be not a whymsy I know not but feare much it is."

Two books of 1508 were printed on vellum, Fisher's *Sermon on the Fruitful Sayings of David*, and Richard of Hampole's *Devout Meditacions*, and there are copies of both in the British Museum.

The year 1509 was a very important one in De Worde's life. The death of Henry VII was very soon followed by that of Margaret, Countess of Richmond and Derby, De Worde's special patroness. However, the output of books this year was the largest of any year of his life. The royal funerals and the coronation would no doubt attract large crowds to London and so encourage business. Out of some thirty books printed this year I can only trace five as having been issued before Henry VII's death, that is roughly during the first four months of the year. To one of these I should like to draw your particular attention. It is called *Nicodemus Gospel* and there is a copy in the University Library. Its colophon runs " Enprynted at London in Flete strete at the sygne of the sonne by wynkyn de worde, printer unto the moost excellent pryncesse my lady the Kynges moder. In the yere of our lorde god

MCCCCC. ix. the xxiii. daye of Marche." This colophon and that to the *Golden Legend* of 1498 are two of the proofs we possess that De Worde began his year on January 1 and not as was very common on March 25. The book you will notice was issued on March 23, 1509, and Henry VII and his mother the Countess of Richmond are both referred to as alive. Had De Worde begun his year on March 25 and meant by March 23, 1509, March 23, 1510, both these persons would have been dead. The point is interesting because the custom of printers varied, and, as I shall show later, De Worde's contemporary Julian Notary dated the other way. In a few books printed between the death of Henry VII and that of the Countess of Richmond and Derby De Worde calls himself printer to the King's grandmother.

About this time De Worde had a second shop in St Paul's Churchyard with the sign of Our Lady of Pity, but he does not appear to have kept it long and it is only mentioned in some colophons of this year. It was perhaps in reference to this sign that in some of his books he placed at the end a small cut of Our Lady of Pity in place of his usual device.

The books issued in 1509 were of all kinds. Funeral sermons on Henry VII and the Countess of Richmond, congratulatory poems addressed to Henry VIII, and a very large number of popular poems and stories, among them such books as *Richard Cœur de Lion*, *The Conversion of Swearers*, *The Fifteen Joys of Marriage*, *The Parliament of Devils*, and Hawes's *Pastime of Pleasure*. Perhaps the most interesting of all is the edition of Henry Watson's version of the *Ship of Fools*. Only one copy of this beautiful book is known, which printed on vellum and is preserved in the Bibliothèque Nationale. Henry Watson was an assistant of De Worde's and acted partly as printer and partly as translator. Another book of this year is worth noticing, the first edition of the York *Manual*. Though it has a full and clear colophon stating that it was printed by De Worde, in fact the statement is repeated twice over, there can be no doubt that it was printed abroad; the type both of text and music, the illustrations, the device, the peculiar size of the paper all go to prove this.

In 1510 a much smaller number of books, only about eight, were issued. The printer seems to have been taking a rest, for he never in any future year prints so small a number. Two curious books were issued in this year, *King Apolyn of Tyre*, and the *Birth of Merlin*, and in the following year *The Demaundes Joyous*. This delightful though sometimes unedifying little book of riddles contains specimens which I remember having been asked as a child, and I suppose they are asked yet. "How many cows' tails would it take to reach the moon?" with the simple answer "One if it were long enough." "How many sticks go to a crow's nest?" "None, for lack of feet."

In 1512 appeared the *History of Helias, Knight of the Swan*, another book only known from a single copy printed on vellum, now in the library of Mr Hoe of New York. One fact about the book is worth noting for the benefit of future bibliographers and that is that neither in the catalogue of the sale in which it appeared nor, consequently, in Mr Hazlitt's description in his last volume of *Bibliographical Collections and Notes*, is there any reference to the book being printed on vellum.

In 1512 there appeared also the first grammatical work by Whittinton, whose various books became so popular that De Worde sometimes issued as many as four editions of one work in a year. At that time printers had not enough type to admit of their keeping it standing for any length of time, and as labour was very cheap they preferred to reset the type for small editions and not print off a very large edition, which perhaps might lie on their hands for a long time.

So many and so varied were the productions of De Worde's press—he printed between seven and eight hundred known books—that it is hard to pick out any for special notice. In 1515 he issued an edition of the *Way to the Holy Land*, which was reprinted in 1524. In 1516 Capgrave's *Nova Legenda Angliae*, of which a copy on vellum is in the Bibliothèque Nationale. In 1517 we have *Troylus and Cressede*; in 1518 *Oliver of Castile*. In 1519 the *Orchard of Syon*, of which several copies were printed on vellum. In 1521 he issued a book of *Christmas Carolles*, of which there is a fragment in the Bodleian; it contains the well-known carol,

> "The bores head in hande bring I
> With garlands gay and rosemary
> I pray you all synge merely
> Qui estis in convivio."

In 1522 an edition of the *Mirror of Golde for the sinful soul* appeared with De Worde's name in the colophon, but the edition is identical with one issued simultaneously by John Skot. From the type it is clear that Skot was the printer, and an examination of several of De Worde's books about this time shows that they were also produced by Skot.

Erasmus begins about this time to be represented amongst De Worde's books by his *Colloquiorum formulae, De copia verborum, Good manners for children*, and a few others, but the printer does not seem to have printed many editions, which is the more remarkable as the booksellers' accounts of the time point to a large demand for Erasmus' books.

In 1530 appeared the *Book of Songs*, of which there is a copy in the British Museum, and it is the first genuine music-book printed by De Worde. The book contained songs for four and three parts, but the one volume known contains the bass only. From this time onwards most of the printer's work was confined to reprinting earlier editions, and only about one in twenty were new books; a good deal of his work was done by other printers, as can be seen from the type, and in the last few years we find him printing books for other people. For his old apprentice John Byddell he printed four or five books in 1533 and 1534, two editions of Erasmus' *Enchiridion Militis Christiani*, a *Life of Hyldebrande*, and another work.

His last book was curiously enough a little poem, *The Complaint of the too soon maryed*. I do not know whether he considered it applicable to his own case, but his own marriage had taken place more than 55 years before.

De Worde died apparently at the very beginning of 1535, for his will dated June 5, 1534, was proved on the 19th January following. It is much to be regretted that this most valuable document has never been reprinted in full, for both the published abstracts (Herbert's *Ames*, 119, 120, and Plomer's *Abstract of Wills*, 3, 4) omit important names. To each of his apprentices, who are not named, he leaves books to the value of three pounds. To Robert Darby, Robert Maas, John Barbanson, Hector, Simon [Simon Martynson?], John Wislyn, and Alard, a bookbinder, bequests are left, and they are described as servants, that is people who were then working for him, either apprentices who had served their full time or journeymen. John Butler, James Gaver, and John Byddell are each described as "late my servant." Besides these, legacies were left to Henry Pepwell, John Gowghe, and Robert Copland, and "Nowell the bokebinder in Shoo lane." This latter was a certain Noël Havy, a Frenchman who came to England in 1523, married an English wife, became a denizen, and continued in business until after 1550. John Byddell and James Gaver were made executors and continued to live and carry on business in his house.

De Worde's wife had long been dead, and no children, if he had any, seem to have survived him. He ordered his executors to purchase land in or near London which should produce at least twenty shillings a year to be given to St Bride's Church to keep an obit for his soul, and we learn from the Survey of Chantries made in February 1547 that the sum thus expended was thirty-six pounds. Of De Worde's work as a bookbinder we know little. He inherited the binding tools which had belonged to Caxton, and there are several bindings on which these are used, notably, a very fine example in the library of Corpus Christi College, Oxford.

which can with certainty be ascribed to him. Early in the sixteenth century, when panel stamps were generally used, Caxton's dies seem to have been parted with, but we cannot at present point to any panel as having been used by De Worde. It will be noticed, however, that the three binders mentioned in the will, Alard, who was then in his employment, James Gaver, a former assistant, and the independent workman Noël Havy of Shoe Lane were all foreigners, and therefore most probably used their own panels of foreign design.

James Gaver belonged to a well-known family of Low Country binders and would doubtless use panels similar to theirs. When we find early printed English books in their original bindings it is fairly safe to assume these bindings to have been executed in England, for there could have been practically no exportation of such books, and we know of a considerable number stamped with distinctly Netherlandish panels. This James Gaver continued to live with Byddell after De Worde's death at the sign of the Sun and issued one book with his name, an edition of Stanbrydge's *Accidence* from that address in 1539. Whether or no this book was really printed by him is uncertain; for it is identical, with the exception of the colophon, with an edition printed nominally by Nicolas Bourman. At the end of Gaver's edition is a very rude woodcut of the Sun (used as a printer's device and referring to the sign of the shop), which had been used previously by Lawrence Andrewe in his edition of the *Mirrour of the World*.

Gaver had taken out letters of denization on March 2, 1535 in which he is described as James Gaver, stationer, from the dominion of the Emperor. In 1541 he is entered in the Returns of Aliens and pays a tax of £1. 7s. He died in 1545 and his will was proved on June 15. He there calls himself a bookseller and requested that he might be buried in St Bride's Church, before the altar of St Catherine, "neare unto Wynkyn de Worde sometyme my master." The overseer to the will was William Stewarde, a stationer, his son-in-law.

Julian Notary, who had been printing in Westminster since 1497, moved soon after 1500 into St Clement's parish, just outside Temple Bar. He appears to have taken possession of the shop lately vacated by Pynson, which during the latter's tenancy seems to have had no sign. By Notary it was given the sign of the Three Kings. The first dated book from this new address was an edition of the *Golden Legend* issued in February 1504. Just as from the very full colophon of the preceding edition issued by W. de Worde in 1498 it is clearly proved that De Worde began his year on the first of January, so from the colophon of this edition we can prove that Notary did the opposite and began his year on the twenty-fifth of March. It runs as follows. "Whyche werke I dyde accomplysshe and fynysshe at Tempell barr the xvi daye of Feverer. The yere of oure lorde a Thousande ccccciii. And in the xix yere of the reygne of kynge Henry the vii. By me Julyan Notary. Thys emprynted at Temple Barre by me Julyan Notary." Now the 16th of February 19 Henry VII was 1504 in our ordinary computation, so that Notary clearly did not end his year until March 24. This *Golden Legend* contains a very curious and mixed collection of illustrations. Some had belonged to W.

de Worde and some to Caxton. There are besides five metal engravings in the "*manière criblée*," obtained doubtless from abroad, and some curious engraved initial letters which had apparently belonged to André Bocard. It would be exceedingly interesting could we trace the history of these *criblée* cuts and discover whence Notary obtained them. The designs were clearly taken from a set of compositions the work of a Lower Rhine engraver about 1450–60, known as the "master of St Erasmus," of which a complete set may be found in the British Museum. Notary's, however, are not direct copies but probably imitations of some *criblée* prints copied from the original designs. Such sets are at Munich and Berlin, and another of a similar class may be found in the editions of the *Horologium Devotionis*, printed at Cologne by Ulric Zel about 1490, and John Landen, a few years later. Notary's are very exact copies from these, but engraved the reverse way, a common occurrence in copies.

In 1507 Notary issued a very curious book, the first edition of *Nicodemus Gospel*, which I mention here slightly out of its turn as it contains three more *criblée* cuts and is the only other book except the *Chronicles of England* in which they are found. Only one copy of the book is known and its history is interesting. It appeared in the auction catalogue (p. 354) of Richard Smith's books, sold in 1682, bound up in a volume of tracts, and after the sale it disappeared. On the authority of the catalogue it was mentioned by successive bibliographers, but no one had seen it and its whereabouts was unknown. I was beginning to think that it never existed and that the entry in Smith's catalogue was founded on a mistake, when to my delight, on a recent visit to Dublin, I found the identical volume in Archbp. Marsh's library. It had been bought at Smith's sale in 1682 by Bishop Stillingfleet, and the whole of his library of printed books was bought at his death by Marsh and transported to Dublin. As Marsh's library has never had a printed catalogue and has not been much used, its contents are very little known, though I am glad to say a catalogue of the Early English books is about to be printed. Besides the *Golden Legend*, Notary printed in 1504 an edition of the *Chronicles of England*, and a now lost edition of the Sarum *Hymns and Sequences* of which he issued another edition in the following year.

At the beginning of 1508 an edition of the *Scala Perfectionis* appeared, and the copy described by Herbert is similar to the one which came to the University Library with Mr Sandars's collection. Herbert describes his as in an original binding with Notary's mark, and the one bequeathed by Mr Sandars is similarly bound.

In 1510 Notary issued an edition of the *Sermones discipuli* by John Herolt having on the title-page a very interesting imprint which, translated, runs as follows, "These [*i.e.* the Sermons] are to be sold (where they have been printed) at London in the suburb of Temple Bar near the porch of St Clements in the house of Julian Notary, printer and bookseller, carrying on business under the sign of the Three Kings. And they will also be found for sale in St Paul's Churchyard at the same man's little shop [cellula] from which also hangs the same sign of the Three Kings." Apparently a sign was personal property and could be moved by the owner if he moved to another shop. If, however, he became possessed, on the death of the earlier owner, of a shop with a well-known sign, he would retain it in preference to his own. As an example of the first case we find Rastell giving the sign of the Mermaid to three successive places of business. On the other hand when Byddell succeeded De Worde, he did not transfer his own sign Our Lady of Pity, but retained De Worde's well known sign of the Sun.

What Notary was doing between 1510 and 1515 is unknown, but during that period he issued no dated book. One thing, however, is clear. He had given up his printing-office in the Strand and had moved to a house in St Paul's Churchyard with the sign of St Mark, and there he issued in 1515 another edition of the *Chronicles of England*. Though the sign is not mentioned the address is very clear. "Dwelling in Paul's churchyard beside the west door by my lord's palace." In 1516 he was mentioned as living at the sign of St Mark at the west door. This sign, however, did not please him, and by 1518 he had replaced it with the old sign of the Three Kings. That these two signs succeeded each other on the same house and did not refer to two different houses is I think clear from the wording of the following two colophons. "Imprinted in London in Poules chyrche yarde at the weste Doore besyde my lorde of London's palase. At the sygne of saynt Marke"; and "Inprinted in London...in Paules chirche yarde at the weste dore besyde my lorde of london's palayse, at the sign of the thre kynges."

Among Notary's undated books is a very curious little tract called *A merry gest and a true howe Johan*

Splynter made his testament. The only copy known is in the library of Britwell Court. It begins on the verso of the first leaf,

> This Johan Splynter as every man tell can
> Was the Rentgatherer of Delft and Sceydam.

On the title is a curious woodcut which occurs also on another small tract, the *Mery geste of a Sergeaunt and Frere.*

The best known of Notary's books is his edition of the *Shepherdes Calendar*, a curious medley of matter which was first printed in a translation at Paris by Verard in 1503. That translation was made by a Scotchman who knew very little French, so that the result is rather peculiar. The book was revised and printed by Pynson in 1508 and also later by Wynkyn de Worde. In the only known copy of Notary's edition the colophon is mutilated and the date partly destroyed, but as it was printed in Paul's Churchyard at the Three Kings the date may be safely fixed at about 1518.

Notary made use of two panel stamps on his bindings. One contains in the centre a Tudor rose round which run two ribbons supported by angels. On the ribbons is the motto

> Hec rosa virtutis de celo missa sereno
> Eternum florens regia sceptra feret.

In the upper corners are shields with the cross of St George and the arms of the City of London, while below the rose are the binder's mark and initials. The other panel contains an escutcheon bearing the arms of England and France quartered, ensigned with a royal crown and supported by a dragon and greyhound; the arms of St George and the City of London are in the upper corners.

Of these two panels there are two varieties differing in minor details. They may readily be distinguished by the N in the device, which in the rarer variety has the cross stroke the wrong way.

Robert Copland was for many years an assistant to Wynkyn de Worde, and some have suggested even to Caxton himself from the ambiguous use of the word "master" in the prologue to *King Apolyn of Tyre*, where he speaks of himself as "gladly follows ynge the trace of my mayster Caxton." As Copland did not die before 1548 it is very improbable, though not impossible, that he could have worked under Caxton at Westminster. As an assistant to De Worde he translated a considerable number of popular books from the French and edited others, often adding quaint introductions or prefatory verses. About 1514 he appears to have started in business on his own account, for some copies of the *Dying Creature* printed by W. de Worde in that year have his device on the last leaf. This consists of his mark and initials on a shield hanging from a tree and surrounded by a garland of roses and supported by a stag and hind. Round it is the text from Proverbs xxii., "Melius est nomen bonum quam divitie multe." Like many English devices of the time it is copied from a French one. The garland round the shield refers to the sign of his shop the Rose Garland in Fleet Street, though the first book from this address was not issued until later. A small law-book issued about this time, though it contains his name and device was issued from De Worde's house, the Sun. In 1515 he began to work at the Rose Garland and issued a book called the *Justice of Peace*, of which a copy is in the Cambridge University Library. His next dated book was Barclay's *Introductory to French*, and after this there is a gap of seven years. Altogether before 1535 he printed only some twelve books. The explanation of this is, I believe, that his press was largely subsidised by De Worde. Many books issued by De Worde have prefatory or ending verses written by "Robert Copland the book-printer," and most bibliographers have therefore rashly asserted that they must be reprints of editions which he had previously issued. Copland printed entire editions for De Worde and therefore they contain De Worde's name and address, but Copland as their printer added his introduction.

Another point which seems to prove his dependence on De Worde is that after the latter's death in 1535 up to about 1547 we have no trace of Copland printing at all. He continued to translate and revise for others but the only reference to him as a practical printer is to be found in Andrew Borde's *Pryncyples of Astronomye* printed about 1548, in which the author speaks of his *Introduction to Knowledge* as " now a pryntyng at old Robert Copland's the eldest printer of England." As this book was finished and issued by William Copland, it may be presumed that Robert died about 1548.

Henry Pepwell, who worked at the sign of the Trinity in St Paul's Churchyard, was a native of Birmingham. Of his life we know nothing before the year 1518, when he issued an edition of the *Castle of Pleasure*. In 1520 he printed an edition of the

Christiani Hominis Institutum, a translation into Latin verse by Erasmus of a little tract by Colet. In this he made use of a device which had belonged to his predecessor at the sign of the Trinity, a stationer named Henry Jacobi, but with the name Jacobi erased from the block. Between 1518 and 1523 Pepwell printed eight books, and their rarity may be gauged from the fact that two are only known from fragments, four from single copies, and of the remaining two there are in one case two copies, in the other four.

Pepwell must have been a leading member of the trade, for in 1525 he was appointed, together with Lewis Sutton, a warden of the Company of Stationers. He was a friend of Stokeslay, the Bishop of London, and of his agent Thomas Dockwray, who was afterwards the first warden of the new Stationers' Company. In 1531 he issued an edition of Eckius' *Enchiridion locorum communium adversus Lutheranos*, printed for him at Antwerp by Michael Hillenius. Bale in one of his works mentions this book as follows "No lesse myght harrye pepwell in Paules church yearde have out of Michael Hillenius howse at Antwerp at one tyme than a whole complete prynte at the holye request of Stokyslaye. In a short space were they dyspached and a newe prynte in hande, soche tyme as he also commaunded Barlowes dyaloges to be preached of the curates through out all hys dyocese." The existence of this book was for long doubtful until I finally found the title-page in one of the volumes of Bagford's collections and shortly afterwards I found a perfect copy which had belonged to Latimer in the library of Westminster Abbey.

Pepwell as an important bookseller and good Catholic was probably of great assistance to the authorities in their crusade against heretical books. In 1533 Vaughan writes to Cromwell, "The Bishop of London, Stokeslay, has had a servant [Dockwray] in Antwerp this fortnight. If you send for Henry Pepwell, a stationer in Paul's Churchyard, who was often with him, he will tell you his business."

In 1535 he received by the will of W. de Worde a legacy of four pounds in printed books. In 1539 he issued two grammars of Lily for the use of St Paul's School, but though he is clearly stated in the imprint to have been the printer, there is very little doubt that they were printed at Antwerp. The only copies known of these grammars are bound up with some others by Colet in a small volume now in the Pepysian Library bought by Pepys at Richard Smith's sale in 1682.

In October, 1539, Pepwell, accompanied by William Bonham and Henry Tab, was sent by Cromwell to St Albans to inquire about a heretical book which had been issued from the press there. This I believe to have been a hitherto unknown and unique book "*A very declaration of the bond and free will of man*" issued at St Albans without date or name of printer. The printer was however John Herford, who came to London, in custody of the three stationers to be dealt with by Cromwell.

Pepwell died at the beginning of 1541 and his will was proved on the 8th February, William Bonham the printer being one of the supervisors. Two-thirds of his property was left to Ursula his wife and the remaining third to his children, who were all under age and are not mentioned by name, though no doubt the Arthur Pepwell who was afterwards a member of the Stationers' Company was one.

John Skot commenced to print in 1521, and issued on May 17 *the Body of Policy*, of which there is a copy on vellum in the Cambridge University Library. At this time he was living in St Sepulchre's parish without Newgate and printed there six books. The *Mirror of Gold for the Sinful Soul*, the *History of Jacob and his twelve sons*, the *Book of maid Emlyn*, and two law tracts. In these is found his first device, having his mark and initials on a shield surmounted by a helmet and supported by two dragons. He used a fount of narrow black-letter, and it is clear that besides his own books he printed several for De Worde.

His next move was to St Paul's Churchyard, where he printed eight books, but only two are dated, the *Commendations of Matrimony* of 1528 and *Nicodemus Gospel* of 1529. Besides these there were three law tracts, two grammars, and an edition of *Every Man*. At this address he began to use a device, exactly copied down to the misprint in the motto, from one used at Paris by Denis Rosse, but so carelessly cut that both his name and monogram are printed backwards. He also made use of his first device with his monogram inserted on the shield in place of his mark. Either before or after his residence in St Paul's Churchyard, but probably before, he printed an edition of Stanbridge's *Accidence* "Without Bishopsgate in saint Botolphs parish at George Alley gate."

In 1531 John Toy issued an edition of the *Gradus*

comparationum which has Skot's device at the end and was probably printed by him, though his name is not mentioned. By 1537 he had moved to Fauster Lane in St Leonard's parish, but before this he had got into trouble over a book of which apparently no fragment remains. This publication was an outcome of the extraordinary religious troubles connected with the impostures of Elizabeth Barton, the Maid of Kent, which were put down by Cranmer, and the maid and her associates executed in 1533. Amongst the documents connected with the case was the confession of the printer, and Cranmer entered among his notes " to remember that Dr Bokking did put unto Skotte all the Nun's book to print and had five hundred of them when they were printed and the printer two hundred." This perhaps refers to a tract entitled "*A miraculous work of late done at Court of Strete in Kent, published to the devoute people of this tyme for their spiritual consolation by Edward Thwaytes, Gent.*," of which some manuscript copies exist.

At Fauster Lane Skot printed one book the *Rosary*, dated 1537, and five undated books, *The Golden Litany*, *Nicodemus Gospel*, *The Nut-browne Maide*, *The Book of Herbs*, and *The Battle of Agincourt*. After 1537 we have no trace of him.

Several writers have supposed that he may have been the same person as a printer of the same name who appeared at Edinburgh in 1539, and passed an adventurous career in that town and at St Andrews up to about the year 1571. There seems nothing to support the theory beyond the similarity of names, and an examination of the typography of the two printers shows that they must have been different people.

Of John Butler, the next printer to be noticed, next to nothing is known. Ames asserted on the authority of Maurice Johnson, a gentleman to whom he was indebted for many startling pieces of information, that he was a judge of the Common Pleas. There was a John Boteler a judge, but when he was sitting on the bench our John Butler was engaged as a journeyman with Wynkyn de Worde, who in 1535 left by will to "John Butler late my servant as many printed books as shall amounte to the value of vi£ sterling," a very considerable legacy for the time.

His only dated book, an edition of the *Parvulorum institutio ex Stanbrigiana collectione* was issued in 1529. The colophon states that it was printed "for John Butler," but it absolutely agrees typographically with other grammatical tracts printed "by John Butler" though the type of all is apparently identical with that used by John Skot. Besides the one dated book there are eight undated. Five are grammatical tracts of little interest, and the remaining three are *The Jeaste of Sir Gawayne*, *The Doctrynale of good servantes*, and *The Convercyon of Swerers*. It is interesting to notice that of all the books printed by Butler but one copy is known.

The *Expositiones terminorum legum Anglorum* of 1527, attributed by Herbert and others to this printer, is clearly the work of Rastell.

Butler carried on business at the sign of St John Evangelist in Fleet Street and used as a device a small woodcut of St John with the inscription cut upon it

"Initium sancti euangelii secundum Johannem." This cut was not originally intended for a printer's mark, but was one of a series engraved for a *Horae*. Whether or not he was really a practical printer it is impossible to determine, for the words "printed by" are not always to be taken as literally true. Several books of this period are known, identical except as regards the wording of the colophon, which profess to have been printed by different printers; and many books which bear the name of one printer in the colophon can be clearly proved to be the work of another.

John Toy has to be included as a printer in this lecture on the faith of the colophon of one book, but probably, like some early stationers, he was not always strictly truthful and said "printed by" when he meant "printed for." However, in the Bodleian Library there is an edition of the *Gradus comparationum* whose colophon runs, "Imprinted at London in Poules chyrche yard, at the sygne of saynte Nycolas by me John Toye. The yere of our lorde God M.D.XXXI, the XXX day of May." But the book ends with the device of the printer John Skot, and it is probable that he really printed it. In 1534 Toy had an edition of the *Shorter Accidence* of Stanbridge printed for him at Antwerp by Martin de Keyser, and of this there is a copy in the Cambridge University Library. In 1534 Leonard Cox wrote a letter to "The Goodman Toy at the sign of St Nicholas in Pauls Churchyard" relating to the printing of some translations of portions of the Paraphrase of Erasmus, perhaps the Epistle to Titus which appeared some time later. Toy, like many in his business appears to have married the widow of another stationer Nicholas Sutton. Toy died in 1535 and his will is preserved in the Prerogative Court of Canterbury.

Richard Bankes, who began to print in 1523, was the first of a series of printers who lived at "The long shop in the Poultry beside St Mildred's church door at the stocks," and the first book he issued was a collection of stories which he called "goodly and right pleasant," though they hardly merit that description, entitled *The IX Drunkardes*.

This very curious little book, of which the only copy known is in the Bodleian, sets forth the evils of drink as exemplified by various Scriptural characters. It is a translation from the Dutch and is illustrated with a number of foreign woodcuts and borders.

From 1523 to 1526 Bankes printed five dated books and then ceased to work for thirteen years, though he had one book printed for him by Robert Copland in 1528. When he next appears as a printer in 1539 he was living next the White Hart in Fleet Street and was apparently subsidised to print the works of Richard Taverner, an ardent reformer, a religious writer, and M.P. for Liverpool.

In 1540 immediately after Cromwell's death a series of ballads attacking and defending him were written by Thomas Smyth, clerk of the Queen's Council, and William Gray, a servant of Cromwell's. These came before the notice of the Privy Council and the two authors together with Richard Bankes the printer were summoned to appear at eight o'clock in the morning on Sunday, 3rd January 1541. The notice of this trial brings up a new point of great interest to bibliographers.

Copies of these ballads are still extant in the library of the Society of Antiquaries, several with Bankes's full imprint, and in the indictment he is spoken of as "noted to be the printer." When the case came on however Bankes absolutely denied that he had printed them and laid the blame on Robert Redman lately deceased, and Richard Grafton, who confessed he had printed some of them. On January 4 Smyth, Gray and Grafton were committed to the Fleet. This account shows that the colophons of the early printers, especially in the case of small fugitive pieces, are not to be implicitly trusted, and emphasizes the necessity of a careful study of type. Such a study also often shows that some of the smaller printers were probably not printers at all, but had their books printed for them.

Bankes last appears in 1545, when he issued with Richard Lant the Booke of Cookery, of which there is a copy in the Hunterian Museum at Glasgow.

Lawrence Andrewe, who printed in Fleet Street near Fleet Bridge at the sign of the Golden Cross, was a native of Calais and seems to have had some connexion with John of Doesborch, the Antwerp printer, for whom he translated several Dutch books into English, among them the *Valuation of gold and silver*, and a book called *The wonderful shape and nature that our Saviour Christ Jesu hath created in beasts, serpents, fowls*, etc. He only issued two books with a date, two editions of Jerome of Brunswick's *Boke of Distillacyon* : and not only were they printed in the same year, but they were finished on two consecutive days, the 17th and 18th of April. These two editions vary throughout, and the reason for the double issue seems inexplicable.

His undated books comprise editions of Aesop's *Fables*, and the *Mirror of the World*, in folio, the latter being ornamented with a profusion of miscellaneous woodcuts obtained from various sources; and the *Directory of the conscience* and the *Debate and stryfe betwene Somer and Wynter* in quarto. The *Debate* was to be sold "at the signe of seynt John Evangelyst in saynt Martyns parysshe besyde Charynge crosse" presumably by Robert Wyer and may therefore be dated about 1530. In 1529 Andrewe was apparently associated with Peter Treveris in the production of the *Grete Herball*, for some copies contain his device. This device consisted of his mark on a large shield within a frame of pillars and festoons of very florid work. In some of the initials in his books his mark will be found, showing that they were specially engraved for him. About 1534 a certain printer named Leonard Andrewe, who may have been a relation, was an assistant to John Rastell.

Though about all these early printers we have but meagre information, the career of Thomas Godfray is particularly obscure. His name is found in over thirty books; only two contain an address, in each different, and only one a date. The dated book is the *Chaucer* of 1532 printed at London by Thomas Godfray, itself a curious puzzle. It was edited by William Thynne and had a preface by Sir Brian Tuke, and a copy discovered by Henry Bradshaw in the library of Clare College, Cambridge, has the inscription "This preface I sir Bryan Tuke knight wrot at the request of Mr Clarke of the Kechyn then being, tarying for the tyde at Grenewich." Now Leland, the antiquary, distinctly states that this edition was issued by Thomas Berthelet, and as he was a contemporary of Godfray and Berthelet his words cannot be lightly passed over. The title-page border used by Godfray in the *Chaucer* and the *Gift of Constantine* [1534] was used by Berthelet as early as 1535; and he used in the same year another border frame which Godfray employed in the *Introductorie for to lerne French* by Giles Dewes and which was cracked during the printing of that book. Though we have no direct evidence of the fact it would seem as though Godfray's press was subsidized by Berthelet, to a great extent, perhaps, owing to the latter's occupation with official work. Another point to be noticed is that Godfray never mentions any sign, and this is nearly always found to be the case with printers who worked for others and not for themselves. Two of Godfray's books can be dated from outside evidence. *The Gift of Constantine* was issued early in 1534, for on April 1 Marshall, at whose expense it was published, wrote to Cromwell, "I send you two books now finished of the Gift of Constantine. I think there was none ever better set forth for defacing of the pope of Rome." He also writes, "On the book of Constantine I have laid out all the money I can make, and for lack of it cannot fetch the books from the printers." The second book is Christopher Saint-Germain's *Answer to a Letter* which cannot be earlier than 1535, a date found in the book itself. Godfray worked at one time in the Old Bailey and at another, probably later, at Temple Bar, where he would be close to Berthelet. He made no use of any device.

LECTURE VI.

RICHARD PYNSON AND THE LEARNED PRINTERS.

HAVING treated in my last lecture of Wynkyn de Worde and his followers who represent the popular printers of the time, I come to-day to Richard Pynson and his school, who represent the more learned press. We have Haukins and Redman, Pynson's successors; William Faques who preceded and Thomas Berthelet who succeeded him as printer to the King; and the two Rastells, John and William, brother-in-law and nephew to Sir Thomas More, distinguished in the law.

Pynson, who had learned to print in Normandy, came to England some time before 1492 and started as a printer outside Temple Bar, near the church of St Clement Danes, where he continued up to 1500. In 1500, while still living there he along with his servants were the victims of a murderous attack from a riotous gathering headed by one Henry Squire. In his evidence before the Star-Chamber Pynson stated that his servants were so terrified by frequent threats and attacks that they had left him, and his business was suffering in consequence. This is quite likely to have been the cause of his moving, since he would be much safer and better protected within the City. The move was not a distant one for it was to a house which had belonged to the College of St Stephen in Westminster with the sign of the George next to St Dunstan's Church in Fleet Street, at the corner where Chancery Lane runs into it. He was still but a few doors from Temple Bar, but this time on the right side of it. The year 1500 was also an important one to Pynson, for in that year Cardinal Morton died, who had been a valuable patron, and at whose cost the beautiful Sarum *Missal* of 1500 had been produced.

The first book issued by Pynson at his new address was an edition of the Sarum *Directorium Sacerdotum*, of which there is a perfect copy at Ripon, and an imperfect one in the British Museum, which belonged to Cranmer. The colophon sets forth that it was printed "intra barram novi templi" in 1501, so that the date of Pynson's move to St Dunstan's parish is clearly defined as between the publication of the *Book of Cookery* in 1500 and this *Directorium* in 1501.

The only other book issued this year was a small tract describing the escort and reception to be accorded to Katherine of Aragon.

Before speaking further of Pynson I should like to refer to a point which I emphasized when speaking of De Worde and Notary, that is, the method he followed in dating his books. In my former lectures I asserted that he probably began his year on January 1, and for this reason. The beautiful *Missal* printed at the expense of Cardinal Morton was finished on January 10, 1500, and the Cardinal died in September of that year. If Pynson meant by his January 10, 1500, what we should call 1501, then the book would have been issued after the

Cardinal's death, and though the colophon tells us that the book was printed at his command and expense, it has no mention of his being dead.

The book called the *Rule of St Benet* has sometimes been pointed out as a proof of Pynson beginning his year on March 25. The words in the preface run, " We have...caused it to be emprinted by our wel beloved Rycharde Pynson of London printer. The XXII. day of the monethe of January, the yere of oure Lorde M.CCCCC.XVI. and the VIII. yere of the reigne of oure soverayne lorde kynge Henry the VIII." In this case January 22, 1516, is clearly what we should call 1517, but then the words are not those of the printer but of the author, and it is noteworthy that the printer in his colophon gives no date at all. Doubtless a detailed examination of all books printed between January 1, and March 25, would settle the question, but such an examination is at the present time very difficult. To perform such work exactly a very large number of minute tests have to be formulated and applied, and library after library visited, frequently three or four times as new developments are discovered. When the day comes that our best libraries have adequate photographic facilities these questions will be nearer solution. In 1502 Pynson printed nine books, mostly of a liturgical or scholastic nature, the only one deserving of particular notice being the Sarum *Processional* issued in November. Of this book one copy is known printed upon vellum, which is now in the library of St John's College, Oxford.

In 1503 Pynson printed three books, the first English translation of the *Imitatio Christi*, a new

RICHARD PYNSON

edition of the *Directorium Sacerdotum,* and an edition of the *Mirror of the Life of Christ.* This last book I can only give on the authority of the catalogue of Edwards the bookseller, in 1794, for I do not know where any perfect copy is preserved. A slightly imperfect copy however presumably of this edition was sent to me to examine a few years ago. In 1504 the only dated book is a very fine Sarum *Missal,* of which there is a copy in Emmanuel College and a copy on vellum, slightly imperfect, in Manchester. A curious work on natural philosophy by Hieronymus de Sancto Marcho issued in 1505 has Pynson's device on the title, though it seems probable that he was not the printer. The printer, whoever he was, had not a full supply of numerals, and in place of the two 5's in the date he has printed two small black-letter h's as the nearest approach in type to the numeral. The remaining books of this year are of small interest.

In 1506 Pynson issued an edition of the *Principia* of Peregrinus de Lugo for the Oxford bookseller George Castellain. The imprint runs "per Ricardum Pinson cum Solerti cura ac diligentia honestissimi juvenis ac prudentissimi Hugonis Meslier." Nothing whatever is known of this assistant of Pynson's, whose name occurs only in this book. In this same year was issued an edition of the *Kalendar of Shephardes,* which has a curious preface referring to the earlier edition of 1503, which was translated from the French into the Scottish language and printed at Paris by Antoine Verard; and also a Sarum *Manual* of which there are two copies on vellum, one in the library of Corpus Christi College, Cambridge, the other at Stonyhurst. In 1507 two noteworthy books were issued, an edition of the *Golden Legend*, of which the unique (and happily quite perfect) copy is at Lambeth, and a Sarum *Breviary*, of which two copies are known, both printed on vellum and not quite perfect. The copy in the Rylands Library, the one described by successive bibliographers, was in the collections of Ratcliffe and Count MacCarthy but the date had been cut out of the colophon and the book was usually ascribed to 1508. However, four years ago another copy appeared for sale in an auction and was bought by a private collector. In it the colophon was quite perfect and gave the exact date, August 25, 1507.

Some time after May, 1508, on the death of William Faques, Pynson succeeded to the office of Printer to the King; so that he was not as usually stated appointed by Henry VIII. At first an annuity of two pounds was paid him and in 1515 the sum was raised to four pounds. The office of King's Printer though in many ways lucrative was not without its drawbacks. Though he obtained all Government work he was compelled to lay aside his other work until it was finished, which must sometimes have been inconvenient.

The stirring events of 1509 seem to have had no effect on Pynson's work, and while W. de Worde issued some thirty dated books he contented himself with five, and only one was of any importance from a literary point of view. This was Alexander Barclay's translation of Brant's *Ship of Fools*. This fine folio is printed in black-letter and roman type and contains a number of excellent woodcuts. It also contains the full-page cut of Pynson's arms here used for the first time, and probably only lately granted to him on his appointment as King's Printer, which carried with it the title of Esquire. A small work of Savonarola issued on the eighth of September this year is interesting as being the first dated work issued in England printed entirely in Roman letter, though it is not improbable that the undated oration of Peter Gryphus also issued this year may be a month or two earlier.

It is quite evident on examining the various accounts paid to Pynson by the King that a great deal of his official work has absolutely disappeared. In the many cases where definite proclamations, statutes, or similar productions are quoted, hardly any evidence of their existence is now to be found. Of course having once served their purpose they would become obsolete and mere waste, but as many were printed on vellum— in one case we find 400 skins printed, in another 450— it seems impossible that all could have been entirely destroyed. Numbers no doubt found their way into the hands of the bookbinders and were used to line bindings, for which purpose, having one side blank, they were very suitable; and it is from old bindings that the greater number of those extant have been recovered.

In 1511 he issued the *Pylgrymage of Sir Richard Guylforde*, a most interesting book to read, as it gives a vivid description of a journey to the Holy Land and of the death and burial of Sir Richard while there.

Among the books of 1513 was an edition of Lidgate's *Sege and Destruccyon of Troye*, of which he printed several copies on vellum, one of which is in Pepys's collection, as he did also of the *Statutes of War*, which appeared in the same year, though no copies are now known. Another lost book of this year would have had peculiar interest. Ammonius wrote to Erasmus, shortly after Flodden, "Petrus Carmelianus has just published an epitaph on the King of Scots, stuffed full of womanly abuse, which you may soon read printed in Pynson's type." For the next few years there is little of interest to chronicle in his work.

In 1521 was issued Henry VIII's book the *Assertio septem sacramentorum*. As might be expected many copies were printed on vellum and sent round as presents to the princes of Europe, generally containing an inscription in the King's hand. Four copies at least are now in existence, two being in the Vatican. About this book there is a curious story. Montaigne in the journal of his voyage to Italy in 1581 said, " I saw the original of the book that the King of England composed against Luther which he sent about fifty years since to Pope Leo X subscribed of his proper hand with this beautiful Latin distich, also of his hand,

> Anglorum rex Henricus, Leo decime, mittit
> Hoc opus, et fidei testem et amicitiae.

Unfortunately there is still extant among the State papers a letter from Cardinal Wolsey to Henry in which the Cardinal writes, " I do send also unto your highnes the choyse of certyne versis to be written in the booke to be sent to the pope of your owne hande." So much for the royal author.

Between 1522 and 1525 the first edition of Froissart was issued in two folio volumes, and it is one of those books that puzzle the bibliographer as there are many variations and cancels in the different copies. In 1522 appeared Tunstall's *De arte supputandi*, generally described as the first English printed book on arithmetic. A presentation copy on vellum is in the University Library and another imperfect vellum copy in Christ's College. In 1526 appeared an edition of Chaucer to which I think proper attention has not been paid. It is generally described as consisting only of the *Canterbury Tales*, but this is not the case. The book was issued in various parts of which the *Canterbury Tales* is one, and other works were issued in other parts though no library contains a complete set. How many parts really existed I do not know, but it looks as if the intention had been to issue the complete works. In 1526 and 1527 editions of Henry's VIII letters against Martin Luther were issued, but unlike the *Assertio* no copies seem to have been printed on vellum.

Beyond the issue of three small law-books in 1528 Pynson seems to have issued nothing up to his death in 1530, and there is no obvious reason to account for this. There is something rather mysterious about the relations between Pynson and Robert Redman, who will be noticed later, for all the books printed by the latter between 1528 and the death of Pynson, when he succeeded to his shop, bear no address, and it is just possible that some arrangement had been made between them. The bindings produced by Pynson are of very rare occurrence. He used two panels of a small size. One contains his mark within a broad border and is very similar in design to his device. The other contains the Tudor rose in the centre with a border of foliage and flowers and vine-leaves in the corners. There is an example in the British Museum.

Pynson died at the beginning of the year 1530 and his will dated November 1529 was proved on the 18th of February following. He left property in Chancery Lane and Tottenham but there is little of interest in the will itself. He left bequests to his two apprentices John Snowe and Richard Withers on condition of their faithfully serving out their apprenticeships. At the time of his death he had only one child alive, his daughter Margaret, who had married first a certain William Campion, probably a stationer, by whom she had two daughters, Amye and Joane, and secondly a man named Warde. Pynson's son Richard is described as lately deceased, but he had left a daughter Joan who was old enough to be married in 1537. It is almost certainly this son Richard, and not as usually asserted his father, who took out letters of denization in 1513, for Richard Pynson the elder could never have risen to be King's Printer and to have the right to bear arms without having been made a denizen. Everything points to the fact that he was not only denizened but naturalised, but the son who from his age must have been born abroad would require letters of denization also.

After Pynson's death at the beginning of 1530 a certain John Haukins completed and issued the curious book on which Pynson had been at work for some time, *L'Eclarcissement de la langue Française* by John Palsgrave. The history of this book is somewhat mysterious. At the end of 1523 an indenture was made out between John Palsgrave and Pynson for the printing of 60 reams of paper at six and eightpence a ream. Another indenture of the same year was for printing 750 copies of *Lesclarcissement de la lange Francoys*, containing three sundry books. Pynson engaged to print daily a sheet on both sides [that is four pages] and Palsgrave agreed not to keep him waiting for copy.

The final indenture dated January 18, 15 Hen. VIII [1524], between John Palsgrave, prebendary of St Paul's, and Richard Pynson citizen and stationer of London, arranges for the printing of a book named *Lez le Clarissimaunt de la lange Francois*, containing three books, with certain tables and a French vocabulist. Palsgrave will pay six and eightpence for each ream of paper 20 quires; 750 copies are to be printed, of which Pynson shall have as many as, at a price agreed between them, will pay him at the above rate. Clauses to be inserted that Pynson shall not print more than 750 till that number is sold, and that Palsgrave shall deliver the copy from time to time truly corrected. The book as we now have it consists of three parts. The first has the title, introduction, and index, and a privilege from the King dated September 2, 1530, so that the printing must be after that date. The second contains two books, one on pronunciation, the other on the nine parts of speech and ends with Pynson's device. The third part contains the third book with tables of words and ending with this colophon, "Thus endeth this booke called *Lesclarissement de la langue Francoyse*, whiche is very necessarye for all suche as intende to lerne to speke trewe frenche; the imprintyng fynyssed by Johan Haukyns the XVIII daye of July. The yere of our lorde god, M.CCCCC. and XXX." In the preface of the author to the King he speaks of his having formerly written two "sundrie" books on the subject which he had "offred" to the Princess Mary, the King's sister, and Charles Brandon, Duke of Suffolk, her husband. These two books I take to be the two contained in the second part of *Lesclarcissement* which were certainly printed by Pynson. A curious piece of information about the book occurs in a letter dated April 13, 1529, written by Stephen Vaughan to Cromwell. In it he says that " he wishes to learn French and when in London asked Mr Palsgrave for one of his books, which he refused. Requests Cromwell to get one for him, as Palsgrave will not refuse him. Hears he has told Pynson to sell them only to those he names lest his profit as a teacher should be diminished. Would esteem one no less than a jewel and will send Cromwell something of greater value in return."

Now in Pynson's first agreement he was to print 60 reams of paper, that is 1200 quires and 750 copies of the book. This gives to each copy of the book $1\frac{3}{5}$ quires of paper, and taking twenty sheets to the quire, each book would contain 64 leaves. The middle part of *Lesclarcissement* consists of 60 leaves without any title-page or prefatory matter. My theory is that this portion with a title and preface was issued in or soon after 1524. We see that Palsgrave was very chary of selling copies, so that many would remain in the printer's hands. At a much later date the large third part containing the vocabularies was put in hand and finished in July, 1530 by John Haukins. The copies of the middle part were added to it, their old title and prefatory matter cancelled and a new title, index, and prefatory matter printed for the whole book some time after September 1530.

As the wording of the agreements mentioned above is slightly ambiguous, a word of explanation is necessary. The 6s. 8d. which Palsgrave paid per ream was the price of the paper itself and not the printing of it. Palsgrave was to pay for the paper and Pynson was to have so many copies to sell at a rate agreed upon as would repay him his outlay in printing. The average cost at that time for printing in comparison to paper was as four to one. The ream costing 6s. 8d., the printing and printer's profit of the ream would amount to £1. 6s. 8d., thus the total cost of a ream, paper, printing, and all would be £1. 13s. 4d. I have collected many notices of the price of printing-paper from the time of Arnold's *Chronicle* (about 1496) during the whole of our period, and while writing-paper went so high as thirty-four shillings and fourpence a ream, printing-paper was invariably 6s. 8d.

William Faques, who succeeded Peter Actors, the first stationer to the King, as the first King's Printer, is but a very shadowy figure, and of his life we know nothing. He was a native of Normandy, and Herbert suggests, but without any reason, that he may have learned his art with Jean le Bourgeois. The only date connected with his books is 1504, in which year he printed a proclamation on the coinage, the *Statutes of the Nineteenth Year of Henry VII*, and a Latin *Psalter*. This last book shows that unknown or not Faques was a skilful printer, for it is one of the most beautiful books issued from the early English press. The type is sharp and brilliant, the printing in red well done, and each page is surrounded with a chain-like border. On the first leaf is the printer's device, a very uncommon one, consisting of two triangles. On one is the verse (Psalm xxxvii. 16) in black letters on a white ground, "Melius est modicum iusto super divitias peccatorum multas." On the second triangle in white letters on a black ground the text (Prov. xvi. 32), "Melior est patiens viro forti et qui dominat." Thus stopping suddenly, not only in the middle of a verse, but the middle of a word.

In the centre is his monogram transfixed by an arrow. The presence of this arrow is very puzzling, for it also plays an important part in the device of Richard Faques, William's successor. The early printers were very fond, where possible, of introducing punning allusions into their devices, and this arrow may have some connexion with the name Faques.

Of the *Psalter* some six copies are known, and one printed upon vellum is in the library of Emmanuel College.

The *Statutes*, of which only the British Museum copy is known, has a fairly full colophon, "Here endeth the statutes holden at Westmestyr the xxv day of Janiuere in ye xix yere of ye moste nobyll reigne of kynge Henry the VII. Enprynted in London within Seynt Helens be Guillam Faques ye kyng Prynter." St Helens was in Bishopsgate Ward. Two other books with Faques' name are known, an edition of the *Vulgaria Terentii*, and the homily of Origen, *De beata Maria Magdalena*. Both these little books are without date and both are stated to have been printed in Abchurch Lane. Two copies are known of the Origen, of the *Vulgaria* the only known copy is in the library of Corpus Christi College, Cambridge. These books are curious in having the top part of the title cut in wood, a peculiarity which Pynson copied immediately afterwards, perhaps employing the same woodcutter.

William Faques no doubt died in 1508 for R. Pynson was appointed King's Printer in that year. He was succeeded in business by Richard Faques, who was presumably a near relation and a foreigner, and we may discard as a fable the statement made to Ames by "Mr Thomas Wilson of Leeds, in Yorkshire," who in a letter 2 April, 1751, informed him that Richard Fawkes, printer, was second son of John Fawkes, of Farnley Hall, in the said county, Esq.; and in a pedigree he had of that family he was called printer, of London.

In 1509 Richard Faques issued the *Salus corporis salus anime* of Gulielmus de Saliceto, a book printed with his predecessor's type and with the chain ornament, and in 1511 issued with W. de Worde an edition of the Sarum *Missal*. His device, very well cut, is a copy of that used by Thielman Kerver, but with some alterations. Two unicorns standing amid flowers and foliage support a shield hung from a large arrow on which are the initials R.F. and a maiden's head, in reference to the sign of the shop where he carried on business, the Maiden's Head in St Paul's Churchyard. Below on a ribbon his name is cut, Richard Faques. In 1521 when we next find a dated book he had removed to another shop in the churchyard with the sign of the A. B. C. He had also made a change in his device, anglicising his name by cutting out the "ques" of Faques and inserting "kes" in its place in type. It is interesting to notice that in all the five dated books which he issued he made a change in the spelling of his name, each time making it more English. In 1509 it was Fax, in 1511

Faques, in 1521 Fakes, in 1523 Faukes, and in 1530 Fawkes. In 1523 besides his shop in St Paul's Churchyard he had a dwelling-house, where probably he printed, in Durham Rents, in the Strand. In the assessment of Aliens for the subsidy of 1524 as printed by the Huguenot Society a Richard Far is given in the Strand, and I think it is quite likely that Far is a misreading for Fax. A good deal of the Strand was in Westminster, and in the Westminster denization roll of 1544 is one Amelyne Faxe, widow, aged 70 years, in England 55 years. "Hath the Kinge's Magestie's proteccion of his grace gyft to Richard Faxe her husbond, late deceased, to remayn and dwell within this realm, but her landlorde will not suffre her to dwell in house."

The Richard Fawkes here mentioned, if he was the same as the printer, died in 1538 and was buried in the parish of St Martin in the Fields.

The last book issued by R. Faques was the *Mirrour of Our Lady* of 1530 printed at the desire and instance of the abbess and general confessor of the Monastery of Syon. It is a beautiful volume with several woodcuts, one of them signed E. G. similar to one in a Pynson book, and a number of curious initial letters. A certain Michael Fawkes was joined with Robert Copland in 1534–5 in printing an edition of the *Tree and* XII *frutes of the holy goost*, and also printed the *Consolation of timorouse and fearfull consciencys* of which there is a copy in the British Museum, but beyond this nothing is known of him.

Robert Redman began to print in the year 1523, his first book, of which there is a copy in the Cambridge University Library, being an edition of Fitzherbert's

Diversite de courtz. His next dated book, an edition of the *Magna Charta*, was issued in 1525, and in this his address is given at the sign of the George in St Clement's parish, elsewhere described as just outside Temple Bar. Considering the custom of printers in successively occupying the same houses, it is probable that this is the same printing-office as that used first by Pynson and then by Julian Notary. Now at this time Pynson was at work close by on the other side of Temple Bar and his sign was also the George, and when Redman not only used his sign but began to issue editions of the books he had been accustomed to print, we can understand the older printer becoming very indignant. The publication of an edition of Littleton's *Tenures* by Redman apparently brought matters to a crisis, and Pynson in his edition of the same book issued in 1525 gave expression to his feelings in a somewhat strongly worded " letter to the reader." In it he points out how much more correct and well printed his work is as compared with that of Robert Redman, or more properly Rudeman, for among a thousand it would be hard to find one more unskilled. He wonders how he can call himself a printer, unless the devil made him one when he made the cobbler into a skipper. Formerly the scoundrel professed himself as skilled a bookseller as ever came from Utopia, well knowing a thing can be called a book when it has merely the appearance of one and little else. He finishes up by abusing him for daring to promise that he could print the laws of England properly, and asks the reader to judge for himself. To this invective Redman returned no answer but continued to issue his books as before, and though Pynson on one or two other occasions repeated his attacks they produced no effect, unless perhaps the addition to his colophons which Redman sometimes printed, "Si deus nobiscum quis contra nos."

It was suggested by Herbert and others that Redman removed into Fleet Street before April 18, 1527. There certainly is an edition of the *Modus tenendi unum hundredum* of that date with a distinct colophon stating that the book was printed by Redman at the George in St Dunstan's parish that is within Temple Bar, outside being St Clement's parish. But this date must be a misprint, not only because colophons of 1528 again give him as living in St Clement's parish, but also he could hardly have occupied Pynson's house while the latter was still at work. It is a curious point to notice that during the period between 1528 and 1530 Redman gives no address in his books. Immediately on Pynson's death, however, at the beginning of 1530 Redman not only moved into his house, but took over part of his material, and for the future made use of one or other of his old rival's devices. Previous to this he had no distinctive device, but made use of some small cuts, one of the Infant Christ seated, another of St George, and a third of the Trinity. He used in all, three of Pynson's devices, the original black block with the white monogram with which Pynson had first started, a rarer small metal device, not often used, which has a pierced ribbon at the bottom in which the printer's name could be inserted in type, and the large late wood-block. On March 23, 1530, he issued the first book from his new address, an edition of the *Natura Brevium*. The book is dated March 23, 1529, but this must of course mean 1530, and shows that at any rate as regards law-books Redman began his year on March 25. To law-books Redman mainly confined his attention, and the books in other classes which he issued are not as a rule of much interest. He appears not to have had much initiative, but contented himself with reprinting popular books. In 1533 this practice led him into trouble, for in February of that year he was bound over in the sum of 500 marks not to sell the book called '*The division of the Spiritualty and the Temporalty*' nor any other book privileged by the King. The printing of this book had been granted to Berthelet, who as King's Printer would be in a position to enforce his rights. It was of this book that More wrote in his *Apology*, "And in this poynt they lay for a sample the goodlye and godlye, milde and gentle fashion used by him, whosoever he was, that now lately wrote the booke of the division betwene the temporaltie and the spiritualtie, which charitable mild manner they say that if I had used, my woorkes would have been read both of many moe, and with much better will." The authorship though unknown to More is generally ascribed to Christopher St Germain.

Among the more interesting books printed by Redman may be mentioned editions of the *Life of Christ*, *The Frute of Redempcion*, *The Pomander of Prayer*, Fewterer's *Myrrour of Christes Passion* written in 1533 and printed the year after, and Whitford's *Dayly Exercise* in which he complains, like Caxton, that having been asked to write out his book over and over again he had thought better to print it and thus save himself so much labour.

A very curious border piece was sometimes used by Redman, as for instance in the English translation of Lyndewode's *Constitutions* of 1534 and the *Book of Justices of Peas*, which contains in the lower margin the initials I. N. and I. M. Who these initials refer to I have not been able to discover, but the design of the border was popular and was used by Pynson and at Antwerp by Michael Hillenius, while it is also found in some of Tindale's books. It may have been engraved for some stationer and afterwards obtained by Redman. The last important work on which Redman was engaged was a folio edition of the *Bible*, which he printed in partnership with Thomas Petyt for Berthelet.

Redman died in 1540 between October 21, the date of his will, and November 4, when it was proved. He left his property to be divided into three parts. The first for bequests and funeral expenses, the second to his wife, and the third to his children. One of his executors was his son-in-law Henry Smith, a stationer and printer of law-books who lived at the sign of the Trinity, without Temple Bar, in St Clement's parish, perhaps the very house which had once been in the occupation of Redman, who had used a device of the Trinity as one of his early marks. Redman's wife Elizabeth, whose maiden name had been Pickering, continued to carry on the business by herself for a short while, but retired on her remarriage with Ralph Cholmondeley, when the printing-office and its effects passed to William Middleton.

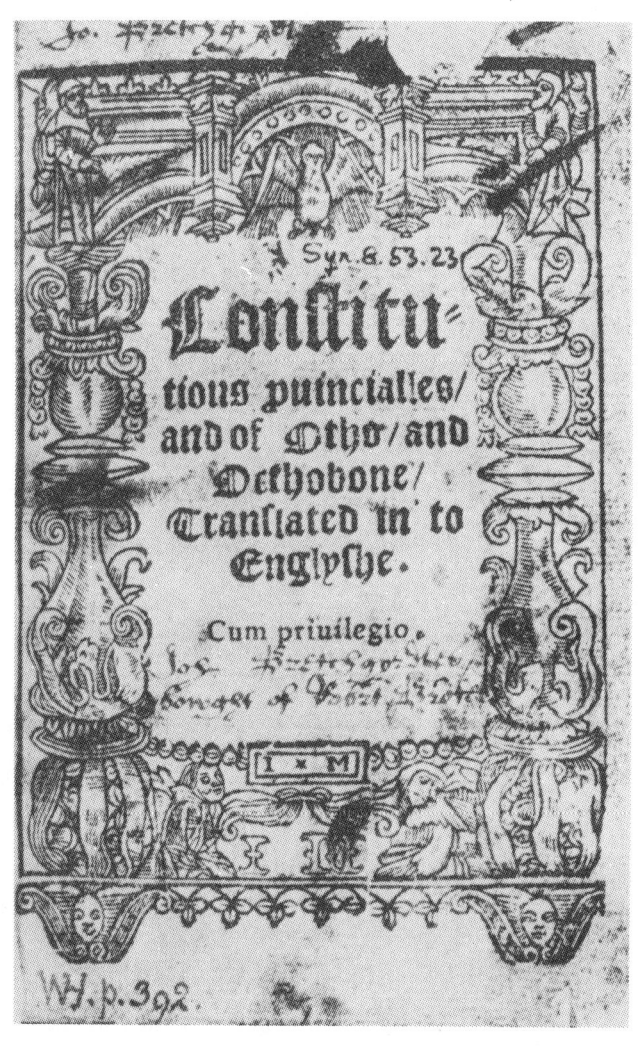

TITLE-PAGE TO R. REDMAN'S EDITION OF LYNDEWODE'S 'CONSTITUTIONS' OF 1534.

A certain John Redman, born in 1508 and who was in business as a stationer at least as early as 1530, may have been a relation, though there is no direct proof. He printed at Southwark a small work of Cicero for Robert Redman which lends some probability to the theory, and on Robert's death in 1540 he appears to have moved to London to a shop in Paternoster Row with the sign of Our Lady of Pity.

Thomas Berthelet, Pynson's successor as King's Printer, seems to have been at one time in his employment as apprentice or assistant, and may most probably be identical with the Thomas Bercula or Berclaeus who speaks of himself as the printer in several books issued by Pynson. The earliest in which this name appears is an edition of the *Vulgaria* of Whitinton issued in 1520, and after the editor's preface is a short address to the reader by Thomas Bercula Typographus. There is no definite statement to connect Bercula and Berthelet and yet it is hard to see who else the name could apply to. Berthelet may have come into the business to take the place of Pynson's son Richard who had died, but it is curious that there is no reference to him in Pynson's will. In 1524 Berthelet married his first wife, for it seems most probable that he is the person referred to in the following entry in the register of marriage licences granted by the Bishop of London, "1524 August 23 Thomas Barthelett of St Dunstan in the West and Agnes Langwyth, widow, at St Bride's, Fleet Street." In 1528 he started business on his own account, his first book being Thomas Paynell's translation of the *Regimen sanitatis Salerni*: 'This boke techying al people to governe them in helthe.' The colophon runs "Imprinted in London in Flete Strete in the House of Thomas Berthelet nere to ye cundite at ye signe of Lucrece. Anno domini 1528, mense Augusto." Another work of Paynell's, entitled *The Assault and Conquest of Heaven*, was issued in 1529. Berthelet's shop must have been further east than Pynson's and close by Wynkyn de Worde's, though it is impossible to say on which side of Fleet Street it stood. Two other early books often quoted as earlier than 1530 may be noted here. One is a work by Wakefield on the divorce controversy, quoted by Wood in the *Athenae Oxonienses* and dated by him 1528, the other is an edition of the *Statutes*, the first book of Berthelet's given by Herbert and said to be dated 1529. The first of these, of which there is a copy in the Bodleian, is without date, but from the subject-matter cannot be before 1533 and from the fact that the printer is styled King's Printer could not be earlier than 1530. For this latter reason also the *Statutes* of 1529 must be non-existent. One or two books, such as Erasmus on the Lord's Prayer, in which he does not style himself King's Printer, may perhaps be assigned to before 1530.

On February 15, 1530, immediately after Pynson's death, Berthelet was appointed Printer to the King with an annuity of four pounds. Just a few days before he had issued an edition of Paynell's translation of the *Regimen sanitatis Salerni* which is very interesting as illustrating a point I have several times referred to. The book is dated February, 1530, and Berthelet is not spoken of as King's Printer, a fact he emphasized very particularly as soon as he had risen to that position. We may conclude therefore that in this colophon February 1530 does not mean February 1531 and that therefore Berthelet calculated his year as beginning on January 1.

After the Royal appointment Berthelet's press started on an active career, and the first book issued, naturally one on the political question then engaging the attention of Europe, was an important work on the subject of the royal divorce. This was the first edition, in Latin, of the *Determinations of the most famous Universities*, written and collected for the purpose of strengthening the King's position. It was issued in April and an English translation appeared in the following November. In June two proclamations were issued, one for the punishing of "vagabondes and sturdy beggars," the other for "dampning" erroneous books and heresies, and prohibiting the translation of Scripture. The first of these is printed on one sheet of paper, the second on a sheet and a half, and the printer was paid for 1600 of the two £8. 6*s*. 8*d*., that is, one penny each for the single sheet and a penny halfpenny for the sheet and a half. In each penny the cost of the paper represented one-fifth, for "paper royall" which was then used for printing cost six and eightpence a ream, the remaining four-fifths representing all the cost of printing and the profit.

The number of proclamations printed by Berthelet now in existence is very large, but nothing to compare with the number which he must really have printed. Looking at such documents as his three year's accounts which have been preserved, and casual entries in records, it is clear that not a tithe is represented by existing specimens, though we are more lucky in his case than in Pynson's, almost every one of whose productions in this class has been destroyed.

In 1531 Berthelet issued the first edition of Sir Thomas Elyot's *Book named the Governour*, a work he frequently reprinted, and he seems to have published all Elyot's works, which with their various editions amount to a considerable number. In 1533 he issued the *Dialogue betwixte two englyshe men, wherof one was called Salem and the other Bizance*. This work written by Christopher St Germain was an answer to Sir Thomas More's *Apology*, which in its turn was an attack upon St Germain's *Division of the Spiritualty and Temporalty*.

More's answer to the first mentioned book was printed by William Rastell in the same year and entitled *The Debellacyon of Salem and Bizance*.

Among the many beautiful borders which Berthelet used for his books there is one about which a word of warning should be said. It has engraved upon it the date 1534, and as it was in use for several years it has given rise to great confusion in the dating of books.

In 1537 was issued the celebrated *Institution of a Christian Man*, called the Bishops' Book in contradistinction to the *Necessary Doctrine and Erudition for any Christian Man*, or King's book, issued in 1543. Both are versions of the same work, but the first has a Preface of the Prelates to Henry VIII, the second has an introduction by the King.

In 1542 Sir Thomas Elyot published his Latin dictionary, and he appears to have issued or intended to issue a copy on vellum, for in the Bodleian are five leaves so printed, consisting of the title, the prologue to

Henry VIII in English, an address to the reader in Latin, and the table of errata. In 1542 he printed another book upon vellum, Lily's *Introduction of the Eight Parts of Speech.*

In 1540 a work by Nicholas Borbonius was issued at Basle in which there is a Latin poem addressed by the learned author to Berthelet and worded in a most laudatory style. There is no doubt that Berthelet's printing and beautiful type were alone in England able to rival the work of the foreign printers.

On the accession of Edward VI in 1547, Berthelet lost his position as King's Printer, and a new one, Richard Grafton, was appointed, a custom now for the first time introduced, as hitherto the appointment had been for life. For the succeeding eight years of his life we lose the familiar "Regius Impressor" of his colophons, and this helps in a small degree in the dating of undated books. He seems, however, to have become much less active after the loss of his privilege and probably left much of the printing business in the hands of his nephew Thomas Powell, who succeeded him, and his servants. The various purchases of land which he made from time to time point to his having built up a considerable fortune, and apart from his purchases in or about London he had estates in Hereford.

Besides being printer to Henry VIII Berthelet was also the Royal bookbinder, and it is hard to speak too highly of his skill and taste in this direction. The beauty of the Italian bindings seems to have greatly struck his fancy and it is supposed that he brought over some workmen from Venice both to work for him and to teach his own men. In his accounts rendered to the King, some of which fortunately have been preserved, he speaks of books decorated in the Venetian manner, and the tools used on them are direct copies of those used by Aldus of Venice and other contemporary Italian binders. He made use of one very distinctive tool, similar to some found on Oriental work, by means of which the design appears upon the leather plain while the whole background is gilt. A good deal of the King's binding was done in velvet or white leather. To show the relative value of books and binding take the following entry, "Item delyuered to the kinges the XXIII day of January [1542] a booke of the psalter in Englishe and Latyne, the price VIII*d*., and a booke entitled Enarracones Evangeliorum Dominicalium, the price XII*d*. and for the gorgious byndyng of them backe to backe IIII*s*. IV*d*." Or again, "Item delyuered unto the kinges highnes the XV day of January [1542] a New Testament in Latyne and a Psalter Englishe and Latyne bounde backe to backe in white leather, gorgiously gilted on the leather, the bookes came to II*s*. the byndynge and arabaske drawyng in golde on the transfile IIII*s*."

It is worth noticing, however, that such tenants of Berthelet as were bookbinders were all Frenchmen.

The date of Berthelet's death has been only surmised and never given correctly by bibliographers. Fortunately he held land from the King, so that at his death an inquest was held by the escheator, and full and interesting details are preserved among the *Inquisitiones post mortem*. From this source we learn that his death took place on September 26th, 1555; that Margaret Berthelet was his second wife, and that his eldest son Edward was born on July 24th, 1553. His will, made two days before his death, was proved on November 9th, and by it he left estates to his widow and his two sons Edward and Anthony and legacies to apprentices, godchildren, and charities. A considerable bequest also went to his nephew Thomas Powell, presumably his sister's son, who succeeded him in business.

We have an account of Berthelet's funeral, preserved in Henry Machyn's diary:

"The sam day at afternone was bered master Barthelet sqwire and prynter unto Kyng Henry; and was bered with pennon and cote-armur, and IIII dosen of skochyons, and II whytt branchys and IIII gylt candyllstykes and mony prestes and clarkes, and mony mornars, and all the craftes of prynters, boke-sellers, and all stassyoners."

Among all the early presses that of Berthelet was preeminent for good workmanship. Though he avoided as far as possible the use of illustrations, all the ornamentation he used was in good taste, and in beauty and variety of type he surpassed all printers of the century.

John Rastell stands quite apart from other early English printers. Nearly all we know about them comes from the books they printed while we know little or nothing about their lives. In the case of Rastell we know a good deal about his career, but little about his books. He is said to have been born in London, and was educated at Oxford, afterwards entering Lincoln's Inn and practising the law, in which he was very successful. He must have been of considerable social position, for he married Elizabeth More, the sister of Sir Thomas More. Some time before 1516 he printed an edition of the *Liber Assisarum,* in which he refers to the projected publication of Fitzherbert's *Great Abridgement,* which appeared in three majestic volumes in that year.

About 1520 he moved his printing establishment to a house, "next Paul's gate," which he named the Mermaid, the same sign that he had previously used elsewhere, and a lawsuit which took place about 1534 in connexion with this house throws considerable light on the printer's habits. He appears to have left all practical work to his assistants, going off to his house in the country for months at a time and subletting part of the printing-office to other tenants, among whom were successively William Bonham, John Heron, Thomas Kele, and John Gough, all of them stationers.

Up to the year 1526 Rastell had issued only four dated books, all connected with the law, but in that year he started out in an entirely new style and published two extraordinary books, *The merry jests of the widow Edith,* and the *Hundred mery tales,* neither such as we should have expected from so grave a printer. About 1529 appeared his *Pastime of People,* remarkable for a number of large, clumsy woodcuts. Much has been written about this book and its variations, and it is generally asserted that the British Museum copy is the only perfect one known. There is, however, a very fine and perfect copy in a private library in this country. Another curious book he issued is Lucian's *Necromantia,* of which there is a copy at Shirburn Castle.

In 1530 Rastell was drawn into the religious controversies then becoming violent, and wrote and printed his *New Boke of Purgatory* in defence of the Romish doctrine. This was answered by John Fryth in his

Disputation of Purgatory. Several controversial pamphlets were written by the two opponents with the result that Rastell became a convert to the Protestant religion. This change appears to have been a cause of trouble to Rastell, who writing to Cromwell in 1536 laments the loss of both business and friends. His law earnings, which had been over twenty nobles a term, had fallen to under forty shillings a year and his printing business had fallen off proportionately, and there is no book definitely known to exist dated later than 1530. While most of Rastell's books are legal in character, he is also the printer of several curious interludes or plays such as the *Interlude of the four elements*, the *Interlude of women*, *Play concerning Lucretia*, Skelton's *Magnificence*, Heywood's *Gentleness and Nobility*. The reason for his printing these is not far to seek. He was extremely fond of giving performances of plays at his house and the records have been preserved of a curious lawsuit brought against him by a theatrical costumier on account of dresses supplied. Rastell in 1536 freely expressed his opinions against the paying of tithes, and perhaps on this account or some other not now known was thrown into prison, where he shortly after died. His will dated April 20 and proved October 12 is a rather remarkable document. He had but little to leave. His house which had been made over to his wife on their marriage he bequeaths to her. To his eldest son William only forty shillings, and to his other son John a small annuity he had been left by his grandmother. Other small sums were left to Cromwell and the Lord Chancellor, and for one of his two executors he nominated the King himself, who very naturally renounced probate.

John Rastell's son William was born about 1508 and went to Oxford in 1525, where though according to Wood he studied diligently he took no degree. His first book, issued in 1530, while his father was still printing, was an edition of Caesar's *Commentaries* in Latin and English. His books, like his father's, may be divided into three distinct classes, legal treatises, controversial works mostly by More, and plays and interludes. Amongst the latter are Heywood's *Play of Love, The Pardoner and the Friar,* the *Play of the Weather,* and *Johan Johan the husband and Tib the wife,* copies of them all being in the library of Magdalene College, Cambridge. Another play, Medwall's *Interlude of Nature,* is in the British Museum. While still a printer he appears to have studied law, having been admitted a student in 1532, while he was called to the bar in 1539, where he practised with considerable success. Unlike his father he remained through life a staunch Catholic, and when Edward VI succeeded to the throne, he with many others of his faith sought refuge abroad. He lived in Louvain until Mary came to the throne when he returned to England and was rapidly advanced in his profession, until in 1558 he was made a judge in the Queen's Bench, a position which he retained until 1563. Shortly afterwards he returned to Louvain, where he died on August 27, 1565. After he had given up printing on his own account William Rastell compiled and edited a considerable number of important law-books which continued to be reprinted for a long period. But the work by which he is best known is the edition of the complete works of his uncle Sir Thomas More, published in two volumes folio in 1557 by Richard Tottell.

LECTURE VII.

THE STATIONERS OF LONDON AND THE FOREIGN TRADE.

THE history of the stationers and the book-trade in England during the period between 1500 and 1535 is very obscure and information is difficult to obtain. That the number of stationers was very considerable we know from incidental references, and though references are fairly numerous it is almost impossible in the present state of our knowledge to combine isolated facts into any connected story. One point is quite clear, and every new discovery only tends to emphasize it, and that is that the book-trade with the Continent and the dealings of foreign stationers in this country were infinitely more extensive and important than is usually supposed. While other countries had a perfectly adequate supply of printers and stationers of their own and imported little from their neighbours and practically nothing from England, here the reverse was the case. In the time of Richard III there were but four printers in England, one in Westminster, one in London, one in Oxford, and one in St Alban's, so that we can quite understand the Act of 1484, which was so strong an encouragement to foreigners to trade here in books, and the resulting influx of aliens. There were hardly any restrictions upon foreign trade and there was a ready market for foreign printed books, so that it is not surprising that the more important foreign publishers dealt largely with this country. They appear to have set up stalls and shops in the neighbourhood of St Paul's, where probably the Company of Stationers would not have power to interfere, and to have placed agents in charge of them. Besides this they sent agents round to the various provincial fairs, which at that time were the greatest business centres in the country, with waggon loads of books for sale.

Our sources of information about these stationers are very scattered and also very inadequate. Chief amongst them are of course the vast masses of documents, only partially explored, preserved in the Public Record Office. The list of denizations and the Returns of Aliens have been printed by the Huguenot Society, and these give a little information, though unfortunately the business of the person is often not stated and there is also the further difficulty of his being entered under his correct surname, a name which in the ordinary course of business he rarely made use of himself. The names too are most carelessly spelt and entered. One man for instance occurs under the following names, Frinnorren, Fremorshem, Formishaa, Bringmarshen, and Vrimors. The lists of denizations are of little use as far as our present period is concerned, for from 1509 to 1534 only 240 persons are entered, though in 1535 the large number of 172 was reached owing to pressure being put upon strangers from the Low Countries. In this total of 412 only seven are mentioned as having business connected with the book-trade; in fact in these lists of denizations the occupation of only about one in twenty-five is mentioned. The only returns of aliens before 1535 are the imperfect ones in connexion with the subsidies granted to Henry VIII in 1523-25, and in these the occupations are but very rarely entered. These returns are therefore useless in giving information about otherwise unknown stationers; we can only attempt to trace persons whose names are already known, and this is rendered the more difficult as people are arranged anyhow, according to the ward they inhabited, and there is as yet no index to the book.

Taking all persons residing in England connected with the book-trade, printers, binders, and stationers, from 1476 to 1535, it would not, I think, be far from the mark to state that two-thirds were aliens.

By the Act of 1484 they were exempted from the restrictions imposed on other workmen, and they could by residing close to St Paul's or in the liberties of St Martin's or Blackfriars escape from the jurisdiction of the wardens of the Stationers' Company or the Lord Mayor. Though the immigration of foreigners was always encouraged by Government, it evoked the bitterest hostility from native craftsmen and was the frequent cause of those fights and squabbles which culminated in the famous Mayday riots of 1517. As a case in point may be mentioned the proceedings in the Star Chamber about the end of the fifteenth century, when Richard Pynson sued Harry Squire and others for assaulting and attempting to murder him and his servants. He stated that his assistants were so threatened and assaulted that they had been compelled to leave, and that consequently his business was at a standstill. This perhaps may have been the reason for his leaving St Clement's parish and settling within the City in 1500.

Many aliens in order to obtain the privileges of a native took out letters of denization, but these privileges only commenced from the date of the grant and were not retrospective as in the case of a patent of naturalization. Letters of denization allowed a man to hold, but not to inherit lands; nor did they confer any benefit on the children born previous to the date of the grant. In 1512 an Act was passed for levying a subsidy, and it was ordained that every alien made a denizen should be rated like a native, but that aliens not denizened should pay a double rate. In 1515 this was reversed and denizens again compelled to pay a double rate.

In 1523 as a set-off for levying a subsidy Henry VIII gave assent to an Act which ordered that no alien, denizen or not, using any manner of handicraft within the realm should from henceforth take any apprentice except he be born under the King's obedience; that no alien should keep more than two alien journeymen; and that aliens using handicrafts in London and two miles round should be under the search and reformation of the wardens of the handicrafts within the City of London. This Act seems to have been very laxly enforced, and was practically repeated by a decree of the Star Chamber in 1528, which contained the additional clause that no stranger, not being a denizen and who was not a householder before 15th February 1528, should keep house or shop where he should exercise any handicraft. Finally, in 1534 the celebrated Act against foreign printers and binders was passed, whose terms and effects will be noticed later.

From the earliest times London citizens had been forbidden by their oath of freedom from taking any foreign born apprentice, and the Act of 1523 laid a similar prohibition on all aliens, so that the foreign element was slowly and surely eliminated.

The authorities of the City of London were strongly opposed to admitting any foreigners to the freedom of the City. In a letter from Thomas Berthelet to Morisine written in 1540 he writes, "My lorde mayor told me... that but one stranger born was made freeman these 40 years. I fear the answer is somewhat feigned, for instead of one stranger made freeman this 40 years, there have been six or more. Perhaps they do it to show that they esteem not the liberty of London so light as to admit a stranger born so suddenly when the King's natural subjects must do so long and painful a service before they can enjoy it."

The great centre of trade during this period was St Paul's Churchyard, and the shops there were of two kinds. There were the substantial houses round the cathedral, where the printer or stationer could carry on his business and dwell, but clustered in every direction against the very walls of the church were booths and sheds and stalls. These were simply "lock up" shops of one story, many with flat roofs for people to stand on to view processions, and were used by booksellers and such printers as had printing-offices elsewhere. In Strype's life of Parker we have a description of a shop set up at a later date by Day the printer, "Whereupon he got framed a neat handsome shop. It was but little and low and flat-roofed, and leaded like a terrace, railed and posted, fit for men to stand upon in any triumph or show, but could not in any wise hurt and deface the same. This cost him forty or fifty pounds."

Even the most important printers had stalls before their houses, for we read in an account of a lawsuit in 1536 resulting from an attack on some Frenchmen in Fleet Street, that one of them, endeavouring to escape, concealed himself under the King's Printer's (*i.e.* Berthelet's) stall. Thomas Symonds, a stationer, who was a witness in a lawsuit in 1514, speaks of himself as standing before his stall at seven in the morning, not so very early an hour at a period when the Privy Council assembled at eight.

Though London was the head-quarters of the trade, the stationers did not confine their attention to the City, but travelled about the country attending the various fairs, where so far as we can judge from such few early accounts as are still preserved, a very considerable portion of their trade was done. Great fairs such as that of Sturbridge brought booksellers from far and near, and so great was their fame as a bookselling centre that even at the end of the seventeenth century the great London dealers sent down vast consignments of books which were sold by auction by the leading London auctioneers. All the more important stationers too paid frequent visits to the Continent and attended the great fair at Frankfurt, the principal opportunity for seeing all the latest publications and the recognised time and place for the transaction of business.

In treating of these early stationers I propose first of all to take those who lived and carried on business solely in London, and then to pass on to those who traded both in London and on the Continent. Of very many we know little but the name, and these I can but pass over, touching only on those who are known as publishers of books, or about whom we have some definite information.

The first of these is a certain John Boudins, or Baldwin as he called himself in English, who lived in the parish of St Clement's, Eastcheap. We know of him from one book, an edition of the Sarum *Expositio Hymnorum et Sequentiarum* which was printed for him at Paris by André Bocard in 1502, and which is apparently the first edition with the preface of Badius Ascensius. Boudins died shortly after the publication of the book, for his will dated October 11, 1501, was proved March 30, 1503. He was a native of the Low Countries.

A certain Andrew Rue, who had succeeded his brother John, who had died in 1493, was a stationer like his brother in St Paul's Churchyard. He died in 1517 leaving legacies to relations in Frankfort and others. To Thomas Wallis, priest of St Faith's, he leaves a bound copy of the book of sermons called *Dormi Secure*, and to David Owen of the same church a copy of Quentin's sermons. Two of the executors of his will were John Reynes and Joyce Pelgrim, both to be noticed shortly.

Richard Nele is mentioned as a stationer in a document of 1525, when he petitioned to be transferred to the Company of Ironmongers.

John Taverner, another stationer, was paid £4 in 1521 for binding the books for use in the Chapel Royal.

He died in 1531 and his will was proved in November of that year. In the year following John Sedley, Warden of the Craft of Stationers, died.

It would be useless here, as well as extremely tedious, to go on enumerating the names of stationers of whom we know so little, so I will pass on to some about whom we possess some more definite information.

Joyce Pelgrim, a native of the Low Countries, was early settled in London and in 1504 issued the first book specially printed for him, an edition of the *Ortus Vocabulorum* printed at Paris by Jean Barbier, who had himself but lately left England. This book is only known from some fragments preserved in a binding in Lord Crawford's library. Before 1506 he entered into partnership with Henry Jacobi, and the two, assisted with money by a wealthy merchant, William Bretton, a grocer and member of the Staple at Calais, issued several books connected with the service of the Church. The first and most important was an edition of the *Constitutiones Provinciales* of William Lindewode, issued some time after May, 1506. It was printed at Paris by Wolfgang Hopyl in a most ornate manner and the title-page is one of the most beautiful to be found in an early book, the combination of small woodcuts, woodcut borders, printer's devices, and red and black printing forming a most rich and harmonious whole. Two service books, a Sarum *Horae* and a *Psalterium cum Hymnis*, followed at the beginning of 1507. In 1510 for the same patron they issued the *Pupilla Oculi* of Joannes de Burgo and the *Speculum Spiritualium* of Richard of Hampole printed by Hopyl, as well as another *Horae* printed by Kerver.

On their own account the stationers did little, issuing in 1507 and 1508 three school books. Pelgrim, after the publication of the *Ortus Vocabulorum* of 1504 does not seem to have issued any books on his own account, but he was always connected with William Bretton, and acted as his agent in 1514. He lived in Paul's Churchyard at the sign of St Anne, though nothing is heard of the shop after 1506. It may be that it was one of the many swept away to make room for the new schools built by Colet, for we know that the Trinity, where Pelgrim's partner Jacobi lived, was close by.

Henry Jacobi, probably a Frenchman, was apparently a much more important stationer than Pelgrim and some things point to his having been a citizen of London. He is first mentioned in 1504 as a stationer, and in 1505, as we learn from a note in a MS. in the British Museum, he supplied a large number of books to the King. Shortly after this he joined as a partner with Pelgrim and together they printed books for William Bretton. In 1509 he appears to have separated from his partner, for in the colophon of the *Ortus Vocabulorum* printed by Pynson his name is found alone. In 1510 he went over to France and his name is found in the colophons of some small tracts of Savonarola, printed in Paris in 1510 and 1511 by Badius Ascensius. In 1512 he was back in London, and issued an edition of the Sarum *Diurnale* printed for him at Paris by Hopyl. Francis Byrckman also appears to have been concerned in the publication of this book for his device is printed at the end. About this time Jacobi issued a *Legenda Francisci* printed for him by Jean Barbier at Paris and a *Regula Benediciti* printed by W. de Worde.

Soon after 1512 Jacobi migrated to Oxford, where he opened a shop with his old sign of the Trinity and published there an edition of the *Formalitates* of Antonius Sirectus, printed for him in London by Wynkyn de Worde. A copy of the title-page was found by Mr Proctor in a binding in New College, Oxford, and a copy of the book itself, minus the greater part of the title, was subsequently traced in the British Museum. Jacobi did not long survive his migration to Oxford, but died there in 1514, and on the 11th December administration of his effects was granted to William Bretton through his agent Joyce Pelgrim, while his will was proved the same year in London.

The devices used by Pelgrim and Jacobi were four in number. In the books printed for W. Bretton a large block containing his coat of arms is used. After its first appearance a mistake in the heraldry was discovered and the shield was cut out and a new one inserted in the block, a very early example of what is technically known as plugging. It is curious that this device, though a private coat of arms, was afterwards copied and used as a device by the Paris printer Egidius Gourmont. The device used by Pelgrim and Jacobi together was a blank shield on a ribbon with the motto "Nosce teipsum" with their initials and marks on either side. Jacobi used in the Sirectus a representation of the Trinity with his mark and name in full below on a ribbon. In the books printed for them by Hopyl another device of the Trinity is found.

The bindings produced by the Trinity stationers were of two classes. The folio books were tooled with small dies, the smaller books stamped with panel stamps. Several copies of the *Lyndewode* of 1506 are in existence tooled with identical dies, and one of these copies is lined with unused sheets of the *Ortus Vocabulorum* printed for Pelgrim in 1504. The panel stamps all appear to have belonged to Henry Jacobi, for at present none have been found with Pelgrim's mark. Like many other stationers of the time he used the two panels, having on one side a shield bearing the arms of France and England and supported by a dragon and greyhound and on the other the Tudor rose between two scrolls, supported by angels and containing these verses,

> Hec rosa virtutis de celo missa sereno
> Eternum florens, regia sceptra feret.

Of these he had three series. In the first his initials and mark are quite plain and simple, just as we find them printed on the title-page of the *Lyndewode*. In the second series the ornament was a little more profuse and the mark and initials are omitted under the rose, leaving the space blank. In the last series the ornamentation is more complicated and the initials and mark are so much elaborated that the I of Iacobi looks almost like an A. In conjunction with the second series of panels a square die containing the figure of a dragon is found which had belonged to Caxton and must have been obtained from W. de Worde. Two other panels which belonged to Jacobi are much more foreign in appearance. On one is a figure of Christ seated on a tomb surrounded by the emblems of the

Passion, on the other Our Lady of Pity. Round both runs a legend, and at the end of the second the binder's initials, H.I., occur joined by a knot.

The example of this binding in the Bodleian, very much rubbed and worn, is on some small tracts by Savonarola, two of which were printed for Jacobi in 1510. The binding has on the end fly-leaf an inscription saying that the book was bought by John Yonge, "Aco= nensis," in 1510, and this must be after September 10 of that year, when Yonge was appointed Master of St Thomas of Acres. If Jacobi really was in Paris in 1510 and 1511, this book must have been bound there, and perhaps these two small panels were cut abroad for the purpose of stamping these small volumes for which his other panels would have been too large.

Though Jacobi died in 1514 and Pelgrim is not mentioned after this date, the business at the sign of the Trinity still continued. This we know from a unique but unfortunately imperfect book in the Bodleian. It is called "*Donate and accidence for children enprynted at Parys. Anno Domini* 1515." The imprint on the title states that it was to be sold at Paris at the sign of the Striped Ass by Philippus de Couvelance and in St Paul's Churchyard at the sign of St Katherine or the Trinity. The last leaf which might have contained a colophon is wanting. This Philip de Couvelance was apparently the son of Jean de Cowlance, whose device was the striped ass (asinus riguatus) and who was the same person as Joannes Confluentinus or Jean de Coblentz, whose name occurs frequently in the prefaces of early books.

Birckman perhaps may have had something to do

with the sign of the Trinity, for he seems to have been in partnership with Jacobi in 1512 and used the device of the Trinity on some of the service books of English use which he issued. However by 1518 the shop was in the occupation of Henry Pepwell, who has been noticed before.

The sign of the "striped ass" commemorates the exhibition of a zebra, the first seen in France, which was shown at the Saint-Germain fair towards the close of the fifteenth century.

John Reynes, in many ways the best known of these early stationers, was a native of Wageningen in Gueldres and took out letters of denization on June 7, 1510. Though described in the Subsidy rolls of 1523 as a stationer, he appears to have undertaken other kinds of business, for in 1524 we find him supplying cloth and cotton at the funeral of Sir Thomas Lovell. In 1527 he began his business as publisher by issuing a magnificent edition of Higden's *Polycronicon*, printed for him at Southwark by Peter Treveris, which is remarkable for the excellence of the illustrations. Very fortunately too his mark is engraved at the foot of the title-page and thus gives us the only clue by which we can identify his large series of stamped bindings. In this same year, Reynes in partnership with W. de Worde and Ludovicus Suethon, which name is I believe a misspelling for Sutton, commissioned a magnificent edition of the Sarum *Gradual*, which was printed for them at Paris by Nicolas Prevost, and of which the fine copy, formerly Gough's, is in the Bodleian. In 1530 another service book was printed for him abroad, an edition of the *Psalterium cum Hymnis* for the use of

Sarum and York. Ten years later he issued an *Introductorie for to lerne to rede Frenche* written by Giles Duwes, sometime librarian to Henry VIII, and Tutor of the French language to the Princess Mary. The latest book which Reynes issued was a Sarum *Processional* which was printed for him at Antwerp by the widow of Christopher van Ruremond.

Reynes was certainly in his time the most important stationer of foreign birth settled in England, and his goods valued in 1523 at £40. 3s. 4d. had risen in 1541 and 1544 to £100. The only other foreigner of like importance was Arnold Birckman, who, however, was not settled in this country, but only paid it passing visits and carried on business by an agent.

Reynes's chief fame now, however, rests on his bindings, which are the most frequently found and best known amongst all the early English series. The commonest are those ornamented with a broad roll containing his mark and figures of a hound, a falcon, and a bee, with sprays of foliage and flowers. He had also several series of panels. One pair is particularly good. The first represents the baptism of Christ, who stands in the stream while St John, kneeling, pours the water on His head. The other is a spirited picture of St George and the dragon fighting within an enclosure, round which run various animals and huntsmen. Below are the initials I. R. joined by a knot which must stand for John Reynes, as I have found these two panels used in conjunction with his roll on a binding in the library of St John's College, Oxford.

A more ambitious panel contains what is called the "Arma Redemptoris Mundi," the emblems of the Passion displayed heraldically upon a shield with two unicorns as supporters, and two small shields with Reynes's mark and initials. The companion panel is divided into two parts, one containing the shield with the arms of England and France supported by the dragon and greyhound, the other the Tudor rose with the scrolls bearing the usual verses and supported by angels. These contain, besides Reynes's initials and mark, a shield with the arms of the City of London, so that when these were cut he was probably a freeman.

The last pair, which are late in style and were probably only made for him near the end of his career, contain busts of warriors in medallions between renaissance pillars, connected by ornamental arches, and the whole enclosed within an ornamental border. In the centre between the medallions is the binder's mark.

Reynes died at the beginning of 1544 and his will, made April 8th, 1542, was proved on February 26th. It is a long document and contains many points of interest. His two apprentices, Thomas Holwarde and Edward Sutton, are to receive on coming out of their apprenticeship one hundred shillingsworth of books to be valued according to the way that Arnold and John Birckman sell them to the booksellers. Edward Wright and Robert Holder, his assistants, are left ten pounds in books on condition that they work for Lucy Reynes, the widow, for two years and assist her to realise the stock. Money is left to the poor and for a breakfast to the stationers who come to the funeral, and the residue to the widow Lucy Reynes. It is clear from the fact that Reynes had English apprentices, and from the way he speaks of the stationers, that he had been made a freeman and member of the Company in spite of his foreign birth.

Lucy Reynes did not long survive her husband, for her will, dated April 28, 1548, was proved October 25, 1549. She, like her husband, requested to be buried in the Pardon Churchyard near St Paul's.

Another stationer and binder whose work can fortunately be identified is Thomas Symonds, a stationer of St Paul's Churchyard. His name first occurs in 1514 when he was a witness in a lawsuit quoted by Foxe in his *Book of Martyrs*. He was employed by the officials of London Bridge, and in their accounts for 1525–6 is the entry "paid to Thomas Symonds for binding in boards 17 quires in parchment containing 17 accounts of the bridge works, 6s." This original binding is still preserved and is ornamented with a broad roll containing the binder's mark, a castle and portcullis, a fleur-de-lys, a unicorn, a rose, a pomegranate and the royal arms.

Lewis Sutton was apparently a very important stationer, and must I think be identified with the Ludovicus Suethon who in partnership with John Reynes and W. de Worde commissioned the great Sarum *Gradual* printed for them at Paris by Prevost in 1527. In foreign printed books mistakes in names are not uncommon, and the rare Christian name Lewis combined with two such similar sounding names as Suethon and Sutton, renders the suggestion probable. In 1526 he and Henry Pepwell were the two wardens of the Company of Stationers, which shows that he must have held a high position in the trade. In 1534–5 he was defendant along with Richard Draper, warden of the Goldsmiths, in an action brought by John Gough and John Rastell the printer concerning the latter's printing-office. A later notice of him occurs in the letters and papers of Henry VIII under the date of August 12, 1539. "Receipt by Lewys Sutton, bookbinder of London, of 5 marks from William Hatton of Haldenby for lands in Northamptonshire sold to him." The will of a Lewis Sutton described as belonging to the parish of St Michael le Querne, London, and dated 1541, is amongst those preserved in the Prerogative Court of Canterbury, and is probably that of the stationer. There was another Sutton named Nicholas who was a stationer in London and died shortly before 1531, who lived in St Paul's Churchyard and was probably related to the stationers named Toy.

A stamped binding is known having upon it the initials N. S. followed by a tun, apparently a rebus on the name N. Sutton, and may very likely be his work.

Another person who ought to be mentioned, though neither printer nor stationer, is William Marshall, quoted by many writers as both. He was strongly interested in the Reformation, and spent both time and money in procuring the printing of books in support of the movement. He was a friend of Cromwell, who assisted him with money, and there are in the State papers many interesting letters from the one to the other relating to the publication of such works as the *Gift of Constantine* printed by Godfray, and *Erasmus on the Creed* by Redman, issued in 1534; *De veteri et novo Deo*, printed by Byddell, and the *Defence of Peace* by Wyer, both of 1535. Of this latter book he sent twenty-four copies to the monks of the Charterhouse, which they, under the order of their superior, returned, with the exception of one copy which they burnt. Two books, his *Pictures and Ymages* and *The Chrysten Bysshop and Counterfayte Bysshop*, were apparently printed by Gough, though according to Herbert the latter book has a distinct colophon "Emprynted by Wyllyam Marshall." The most beautiful book printed for him was the reformed *Prymer* of 1535, the work of John Byddell. It is much superior to the usual work of the time, and at least three copies are known printed upon vellum. Like William Bretton, Marshall has printed at the end of some of the books he had commissioned a large woodcut representing his coat of arms.

John Growte, stationer and bookbinder, lived in the Blackfriars next the church door. In 1532 and 1533 he commissioned the widow of Thielman Kerver to print for him two editions of the Sarum *Horae*, which were followed by another in 1534. Growte is identical with the mythical Rouen stationer Jean Groyat, an error which has arisen through the misreading of a colophon. An edition of the *Horae* was printed in 1536 at Rouen "per Nicolaum le Roux pro Johanne groyat et Johanne marchant in parochia sancti Macuti ad signum duarum unicornium manente." A hasty glance has assured someone that Groyat and Marchant were partners living at the sign of the two unicorns overlooking the fact that the last word used in the colophon "manente" must apply to Marchant alone Groyat may very well be the proper spelling of Growte's name, for he was a foreigner, as we know from his being entered in the Returns of Aliens.

During the period of which we are now treating

the presses of Paris and Rouen were actively at work producing books intended for the London market. At Paris, Hopyl, Kerver, Petit, Chevallon, Hardouyn, Prevost, Pigouchet, Higman, Rembolt, in fact all the best known printers and stationers were thus employed. At Rouen we find Morin, Caillard, Olivier, Violette, Bernard, Cousin, Richard, and others, and when we consider that many printed twenty or thirty different editions, we can arrive at some idea of the magnitude of the trade. The most important of the Paris printers was Hopyl; his *Lyndewode*, his Sarum *Missals*, and above all his *Antiphoner*, are splendid specimens of work which no English printer of the time could have attempted to rival. Prevost among other books is represented by a magnificent *Gradual*. Chevallon issued the great Sarum *Breviary* of 1531. Pigouchet and Kerver continued the work they had done so well in the previous century, and issued many editions of the Sarum *Horae*, while Higman and Rembolt printed *Missals*.

Among the Rouen printers Martin Morin undoubtedly took the leading position for beautiful work. He and Olivier produced a large number of Sarum *Missals* remarkable for their fine printing. With regard to these missals there is one point to which I should like to draw attention. Both Morin and Olivier possessed a very large initial M for the title Missale, round the middle stroke of which their name is engraved, Morin or Holivier. Now these letters being practically indestructible, passed to their successors and were still used. This at any rate in the case of Olivier has given rise to considerable confusion, for he is quoted as the printer of books, as for instance the York *Missal* of 1530, which were issued long after he was dead. Among other printers and stationers connected with the English trade may be mentioned Macé, Bernard, Cousin, Violette, and Caillard.

Almost the entire French output consisted of liturgical books. Verard however issued three—well, British—books, for two were in Scottish, and we have it on Pynson's authority that no one could understand them. The third was Barclay's translation of Gringore's *Castle of Labour*, printed when Barclay was himself in Paris about 1503.

Besides these a few grammatical tracts were issued, Violette issued an English Donatus and Jacques Cousin a Stanbridge at Rouen.

The series of beautiful liturgical books which came from the presses of Paris and Rouen afford us information also about many visiting foreign stationers. The *Missal* issued in 1500 by Jean du Pré, but generally ascribed through a misreading of the colophon to 1502, speaks of several new prayers just brought over from England by Jean Antoine the stationer. A Terence of 1504, of which the only copy known is in the University Library (picked up from a small old-book shop in Liverpool), and a Sarum *Breviary* of 1507 were printed for several stationers living in London, one of whom was a certain Michael Morin, no doubt a relative of the famous Rouen printer Martin Morin. Another stationer whose name is found in the Terence is John Brachius, but his name occurs nowhere else, and I have not been able to find out anything about him. Guillaume Candos, at whose cost a Sarum *Missal* was printed in 1509 by Pierre Violette, was apparently in England for a time. Another Sarum *Missal* quoted by Herbert, of which I can trace no copy, was printed for W. de Worde and Michael de Paule, both living in London. Jean Richard of Rouen, who printed so many fine Sarum service books, also came over and was one of the parties in a lawsuit in England at the beginning of the century. From another lawsuit we find that Frederick Egmont, an important stationer of London in the fifteenth century, was still in England, after 1500. I think it will be found, as more documents come to light, that all the principal foreign stationers paid visits to this country or kept agents in London. In very many foreign printed service books we find the expression "to be bought of the booksellers in Paul's Churchyard." I do not know whether that is the equivalent of the modern "to be obtained of all booksellers" or whether it only refers to a small clique of foreign stationers.

Of all the foreign stationers who traded in this country none was more important than Francis Regnault. He was the son of an earlier Francis Regnault, and when a young man started as a stationer in London towards the end of the fifteenth century. Though he returned to Paris about 1496 he still continued to keep a shop in London and probably paid frequent visits to this country. When on the death of his father, about 1518, he succeeded to the great printing business, his English sympathies at once showed themselves, and from that time onward he poured out a continuous flood of books for the English market. Between the years 1519 and 1535 by far the larger portion of the service books produced for England came from his presses. The Act of 1534, and perhaps other causes, seem to have hampered his business to a serious degree and in 1536 he addressed a letter to Cromwell on the subject. The writer states that he lived in London forty years ago and since returned to Paris and continued his trade as bookseller in London, and likewise printed missals, breviaries, and Hours of the use of Sarum and other books. That he has entertained at his house in Paris honourable people of London and other towns of England. He understands that the English booksellers wish to prevent him printing such books and to confiscate what he has already printed, though he has never been forbidden to do so, but his books well received. He asks permission to continue to sell the said "usaiges" and other books in London and the neighbourhood, and asks Cromwell to speak on his behalf to the King, the chancellor and others. He adds that if any faults have been found in his books he will amend them. Considering the date of this letter, it seems much more probable that his troubles were caused not so much by the jealousy of the English booksellers as by the falling off in the demand for the class of books he produced.

Besides printing on his own account, Regnault printed also for English printers. In 1534 he printed for Berthelet the *Interpretatio Psalmorum Omnium* of Joannes Campensis, and in 1538 an edition of the English *New Testament* for Grafton and Whitchurch, edited by Coverdale, and intended to supersede the incorrect version printed shortly before at Southwark by James Nicholson.

One of the latest important undertakings on which Regnault was engaged was a folio edition of the English Bible printed for Grafton and Whitchurch and overseen by Coverdale. The work was progressing favourably when pressure was brought to bear upon the French authorities, and the press, type, and the sheets already printed were seized, most of the printed matter being burnt. By the influence of Cromwell the type seems to have been recovered and conveyed to London, where the work was finished in 1539. The usual account speaks of the printers and presses being brought over as well, but that must, I think, be a little poetical exaggeration, for the presses of England were surely by that time adequate for such work.

In the library of St John's College, Cambridge, is the unique copy on vellum which was printed specially for Cromwell. The titles, woodcuts, and all the initials throughout the volume are beautifully illuminated.

Regnault again attempted to obtain some favour for his books in England, and a letter was written to Cromwell by Coverdale and Grafton on his behalf. Speaking of him they write, " Whereas of long tyme he hath bene an occupier into England more than xl yere, he hath allwayes provyded soche bookes for England as they moost occupied, so that he hath a great nombre at this present in his handes as Prymers in Englishe, Missoles with other soche like : wherof now (by the company of the Booksellers in London) he is utterly forbydden to make sale to the utter undoying of the man. Wherfore most humbly we beseke your lordshippe to be gracious and favourable unto him, that he may have lycence to sell those which he hath done allready, so that hereafter he prynte no moo in the english tong, onlesse he have an english man that is lerned to be his corrector."

From the concluding sentence it would appear that a special attack had been made on these foreign printed service books because of their incorrect printing, and everyone who has examined the very numerous editions of the *Horae ad usum Sarum* printed abroad must admit that any complaints were fully justified. The booksellers evidently succeeded in their opposition, for no more Paris printed service books came to this country until the reign of Mary.

A solitary fragment, rescued like so many others from the binding of a book and now preserved in the Bodleian, is the only record we possess of the presence in England, nominally at any rate, of one of the most important Paris booksellers of the sixteenth century. The fragment consists of the first and last leaves of a little tract of four leaves entitled, "*Psalterium beate marie virginis cum articulis incarnationis passionis et resurrexionis domini nostri iesu xpi nuper editum.*" Of the liturgical interest of the fragment this is not the place to speak; our interest lies in the colophon, which runs "Imprynted at London in Flete aley the .xxi. daye of October by Simon Voter." This, I take it, can refer to no one but the celebrated printer of that name, but whether the colophon exactly means all that it states is another matter. It would seem unlikely that so important a man should come to London to print so small a tract, and besides it has no appearance of English work. Like many others, he had probably an agent of his own in London and printing the tract in Paris sent it to London for sale. The type in which it is printed, a neat black letter, is what may be called a stock type, that is, it was made by a man who made type-founding his business and supplied various printers. On the title-page is a woodcut of the Assumption of the Virgin which I have seen in no other book, but which I hope may some day be the means of settling the real printer.

While speaking of these French stationers and their work, there is one book which though without name of printer or place was printed in English and intended for the English market and should certainly not be passed over, especially as it is in some ways the most remarkable book of the period. It is entitled *The Passion of our Lord Jesu Christ with the Contemplations*, and was translated from the French in 1508 and printed about the same time. The language is peculiar and uncouth and mixed up with foreign expressions. The book is a very large quarto equal in size to the ordinary small folio of the time, printed in double columns. But the extraordinary point about the book is the illustrations, large full-page woodcuts some twenty in number and obviously and unmistakeably German. The style of woodcutting, the dresses, the high-gabled houses, the storks nesting on the chimney-tops are all distinctive. This last gave the immediate suggestion of Strasburg and a very little search showed that the illustrations were almost exact copies of a series engraved by Urs Graf and first published in a volume entitled *Passio Jesu Christi* printed in 1506 by Knoblouch at Strasburg.

One of these cuts, the Crucifixion, occurs in the

York *Manual* printed ostensibly by Wynkyn de Worde in 1509 for John Gachet and James Ferrebouc. Although the colophon of the book states emphatically twice over that it was printed in London for De Worde I have always thought it was printed abroad, not only because of this woodcut, but because the type both of text and music is not De Worde's and because the device used in it is one only found in his foreign printed books and therefore presumably kept abroad.

Much seems to point to Paris as the place of production of both books, and a further piece of evidence exists in the fact that these cuts are found in an edition of *Les Exposicions des epistres et evangiles* printed for Verard in 1511–12. I wrote on the subject of the English book to M. Delisle, who very kindly made some enquiries at the Bibliothèque Nationale, but at present no French original can be traced nor could the Keeper of the Prints throw any further light on the history of the cuts. The translation would appear to have been done in England as it was done at the command of Henry VII. The only perfect copy known is in the Bodleian, but there are fragments in bindings in several libraries, among others Westminster Abbey and Corpus Christi College, Cambridge.

While a serious blow was struck at the French book trade by the Act of 1534, which attacked the supply, Henry VIII himself did it much more serious damage by destroying the demand. Almost the whole of the trade done by the French booksellers consisted in the production and supply of service books; and the great change which was taking place in the services was gradually tending to do away with their use. After this time we find no more editions of the old *Breviary* or the *Missal* produced at all, except for a short time in Mary's reign, and these two books had been mainly produced in France. The *Horae* or *Primer* still continued in use, but its character was so much changed and so much English introduced into it that it was as well produced in England. By this time too the taste for the elaborate borders and illustrations and sumptuous printing which marked the earlier work seems to have disappeared, so that as a general rule the editions sent over from abroad were as inartistic and badly printed as our own.

The competition of the Low Countries must also have had something to do with the decay of French trade, for the Antwerp stationers boasted they could undersell anyone, and certainly such later service books as were printed abroad were printed at Antwerp.

Be the explanation what it may, the fact is clear that the date which saw the passing of the Act of 1534 concerning books, saw the end of French book production for the English market.

LECTURE VIII.

THE STATIONERS FROM ANTWERP AND THE BOOKBINDERS.

THOUGH in regard to books printed for the English market France is most important in numerical strength, it ranks far below the Low Countries in point of interest. The productions of Paris and Rouen consisted almost entirely of liturgical books and a few dictionaries and grammars. From Antwerp on the other hand from the time of Gerard Leeu in the fifteenth century onwards there was an almost unbroken output of books in the English language. Leeu began with reprints of some of Caxton's books, Adrian van Berghen, the printer of *Arnold's Chronicle*, followed, then came Jan van Doesborch with his many curious little English books, and then, most interesting of all, the various editions of the English *Testament* and the multitude of controversial works caused by the religious dissensions in England.

Of Gerard Leeu, "an exceedingly pleasant person" as Erasmus called him, and his English books, I gave some acount in an earlier Lecture; so I must begin the account of the sixteenth century with Adrian van Berghen, Adryan of Barrowe as he calls himself in his English imprints. He apparently began to print about the close of the fifteenth century, living in a shop with the sign of the Great Golden Mortar in the Market; in 1507 he had moved to the Corn Market behind the Town Hall, and at a later date he settled in a house by the Cammerpoort Bridge with the sign of the Golden Missal. In 1535 like many another stationer he got into trouble in religious matters and was accused of keeping and selling Lutheran books, of which a number had been found in his house. In vain he protested that he knew nothing about them and that they had been secretly put there by some enemy, for after a lengthy trial he was found guilty and on January 3, 1536, was sentenced to leave the state of Antwerp within three days, and perform a pilgrimage to, of all places in the world, Nicosia in Cyprus. Nothing further is known of him. But Adrian of Barrowe will ever be had in grateful remembrance as the printer of *Arnold's Chronicle*, for in this he first published to the world the beautiful ballad of the " Nut-browne Maid."

A description of *Arnold's Chronicle* is a difficult matter to undertake. It was one of the books which served as weapons to the late Professor Chandler in his campaign against the subject catalogue of the Bodleian. It contains a list of the London sheriffs, and a guide to writing business letters, a list of the London churches, recipes for making ink, and a variety of other miscellaneous information. And in the middle, apropos of nothing, " The Nut-browne Maid." Chronicle is a misleading title, for the work is really a commonplace book and a very interesting one. Fortunately this and the later edition printed about 1520 by Peter

Treveris are fairly common books and available in a considerable number of libraries. There is a copy in the Parker collection in the library of Corpus Christi College, Cambridge, interesting to the typographer from an accidental cause. The printer when inking the forme accidentally pulled out a type which has fallen sideways upon the top of the rest and impressed on the page an exact full-length image of itself. From the few early examples of such faults as are known it has been possible to obtain a good deal of information as to the sizes and shapes of early types.

Adrian van Berghen printed also an English *Almanack* for 1529 and an edition of Holt's *Lac Puerorum*. This latter is known only from fragments, and some of these have a curious connexion with Cambridge. Among a parcel of unidentified fragments in the Bodleian I came across some pieces of this book which had at one time formed the pad or boards of a binding. From the well indented marks on the paper it was easy to identify the panel which had been stamped upon the binding as the picture of St Nicholas raising the three children, with the name underneath, Nicolas Spiernick, used by Nicolas Speryng, the Cambridge stationer. But the interesting point about the fragments was that on one of them was written the name of N. Speyrinck, thus connecting the Spiernick of the binding with the name as used at Cambridge, Speryng. It is perhaps not strictly right to have included Adrian van Berghen, since he had not so far as we know any personal connexion with England; but the interest of one at least of the English books which he printed must serve as an excuse.

The next printer to be considered is Francis Birckman, the first of the numerous bookselling family of that name, and himself a very important stationer. He was a native of Cologne, who had his headquarters at Antwerp while he carried on business in several other towns, and a great part of his trade was with England. Panzer in his *Annales Typographici* does not mention him before the year 1513, but at an earlier date he was issuing books for the English market and had a shop in St Paul's Churchyard. He is first mentioned in 1504, when in partnership with a certain Gerard Cluen of Amersfoordt in Utrecht he issued a Sarum *Missal*. Beyond the mention of his name in this one book we know nothing of Gerard Cluen, but I think we may take for granted that he was a relation of Birckman's wife, Gertrude van Amersfoordt. Of the *Missal* but two copies are known, one in Trinity College, Dublin, and the other, recently acquired, in the British Museum. Between 1504 and 1510 Birckman's name is not found, but after that date he issued books continuously until about 1528. The history of the Birckman family is very confused, but I cannot help thinking that there were two branches both using the same Christian names. It seems impossible otherwise to account for the large number of books printed in various places almost at the same time, unless they were merely commissioned and Birckman's name put in the imprint. The Birckmans were certainly always travelling about; at a little later date I find one of them within three months at Antwerp, Cologne, London, Oxford, and Cambridge. It is interesting to notice that he had combined his business in Oxford with a visit to young Christopher Froschover, nephew of the printer of the first English Bible of 1535, who was then an undergraduate there.

All the authorities I have been able to consult are unanimous in disagreeing on the history of the family of Birckman, but fortunately at the moment we are only concerned with two, Francis the elder and younger. The elder, whose place of business was in St Paul's Churchyard, must have been an important and rich man and seems to have spared no expense in ornamenting his books. As an example I may mention the great Sarum *Antiphoner* in two folio volumes, of which there is a copy in the Cambridge University Library. The fine title-pages were not the stock-in-trade of the printer Hopyl, but were specially engraved for Birckman, whose mark occurs in them. His Missals also were profusely illustrated, which was not the general rule with such service books.

Francis Birckman was dead before the year 1531, as we learn from the sentence delivered in a lawsuit on April 4, 1531, and registered in the sentence book of the Alderman's Court at Antwerp. A certain John Silverlink had delivered to Birckman a large consignment, over 700 copies, of English New Testaments, for which he was to be paid £28. 17s. 3d. and had only received £3. 7s. 3d. The action was brought by the plaintiff against the guardians of Birckman's children to recover the remainder, which was ordered to be paid minus a sum advanced by Francis Birckman to John van Remonde. His last book appears to have been dated 1529. In 1530 an edition of the Sarum *Processional*, printed at Paris by Prevost, was published in

London with the following colophon, "Venundatur Londonii in edibus junioris Francisci Byrckman apud cimiterium divi Pauli." This Francis was no doubt a son of the earlier Francis, for though his device is different his mark is the same. The device in the *Processional* is the Hen with her Chickens, no doubt referring to the sign of the Birckmans at Cologne and Antwerp of the Fat Hen. This is the only book which I can trace as having been sold in England by the younger Francis Birckman, who probably returned abroad. He does not appear to have printed much, for the business was mainly carried on by Arnold. At a little later date there were certainly two Arnold Birckmans flourishing, one who dealt principally in London, and who had a brother there also in business named John. The other Arnold whose headquarters were at Cologne was dead before 1541, in which year a work of Rupertus, *De victoria verbi Dei* was printed at Louvain for the widow of Arnold Birckman.

John and Arnold were certainly very important stationers in London. John Reynes was the only foreigner who surpassed them in wealth, and when his stock was valued at £100, each of theirs was valued at 100 marks, that is, £66. 13s. 4d. They appear to have acted as foreign agents for many of the London booksellers and to have travelled a great deal from place to place. They did not however always bear the best of characters, for Johann Ulmer writes to Bullinger, "The Byrckmans are careless and by no means to be depended upon, therefore beware"; and in another letter he complains, "The book has not yet reached me, the

Byrckmans are not at all to be trusted." Erasmus was even more abusive on the subject.

Jan van Doesborch, a printer of Antwerp who commenced business some time shortly before 1508, carried on the tradition of printing English books. Of some thirty-two books which he issued more than half are in English and many of them of a very curious nature. He appears to have inherited the business of Roland van den Dorp and his widow, and to have taken on their premises with the sign of the Iron Balance near the Cammerpoort. He is entered in the books of the St Lucas Gilde in 1508, the date of his first dated book, as an illuminator, and this taste is shown in the lavish, if careless, way in which he illustrated his books. Probably the first book he issued was the *Fifteen Tokens*, a little tract describing the signs coming before the Judgement and which he had himself translated from the Dutch. This was followed by three grammars, one of which, Holt's *Lac Puerorum*, has but lately been discovered and is in a private collection. Then come an edition of *Robin Hood*, known from an imperfect copy in Edinburgh, and a fragment of *Eurialus and Lucrece* also at Edinburgh. To the year 1518 may be ascribed the *Life of Virgilius* and a dated book of that year, beginning " This mater treateth of a merchauntes wife that afterwarde went lyke a man and becam a great lorde and was called Frederyke of Jennen." Then we have the *History of Mary of Nemmegen*, an edition of *Tyl Howleglas*, and the story of the *Parson of Kalenborowe*. A last book worthy of mention is one beginning " Of the newe landes and of ye people found by the messengers of the kynge of portyngale named Emanuel.'

Though naturally a much spoken of book as being about the earliest of English books relating to America, really only one leaf of the book is devoted to that subject.

Who the translator of these various books may have been is uncertain. Two were definitely stated to have been done by Lawrence Andrewe, the London printer, and Douce without any apparent reason suggested that others might be the work of Richard Arnold. I think that perhaps several were translated by Andrewe. In some verses appended to his translation of the *Book of Dystyllacyon* he writes,

> After sondry volumes that I dyd deuyse
> As tryfels of myrthe, which were laudable
> Now mynded agayn, my pene to exercyse
> In other maters to the reder more profitable
> And thus abydynge pacyently, for a time seasonable
> My mynde longe vexing with ymaginacyons
> That my work before shold apere more comendable
> Now have I performed a boke of dystyllacyons.

The "sundry trifles of mirth" might well be such books as *Frederick of Jennen*, *The Life of Virgilius*, or *The Parson of Kalenborowe*. The only two books to which he put his name as translator, *The wonderful shape and nature that our Saviour Christ Jesu hath created in beasts, serpents, fowles*, etc., and the *Valuation of gold and silver*, hardly come under the head of "trifles of mirth."

No doubt this large production of English books must have entailed frequent visits to England and a shop or agent there, but the only clue we have to John of Doesborch's residence in this country is afforded by the entry in the lists made for the subsidy of 1523, "Le Johanne van Dwysborow, extraneo, pro xls per annum ijs.," and he is entered in the parish of St Martin in the Fields.

I now come to two printers of Antwerp, in many ways the most interesting of all, but of whom little is known and less has been written. These are Christopher and Hans van Ruremond or Rémonde, who were concerned very specially with the printing and dispersal of the first English New Testaments. Whether or not these two men were related is not very clear, but from the way they were connected in business it seems very probable. Christopher, the more important of the two, was known also as Christopher van Endhoven, which is the name used on his device and under which he is entered in 1524 in the registers of the St Lucas Gilde at Antwerp. Many have thought that Christopher van Ruremond and Christopher van Endhoven were two persons, but as all the books in the two names have the same type and woodcuts, and as in three cases at least we find the device of Christopher van Endhoven and the name of Christopher van Ruremond together, we may take them to be the same. Besides if they were different persons it would be a very marvellous coincidence that they should die on the same date, and their widows simultaneously begin to carry on their businesses.

Christopher began to print in 1523, issuing a Sarum *Manual* and *Processional*; this was followed by four more Sarum service books in 1524 and five in 1525. Until after February 6, 1525, all were printed for Peter Kaetz, the later ones for Francis Birckman. In 1526 he entered on a task which was destined to bring him

DEVICE OF CHRISTOPHER ENDHOVEN.

into considerable trouble, the printing of the *New Testament* in English. In the beginning of the year he printed a Sarum *Breviary*, but for the next year we have nothing else from his press, for a reason to be noted shortly. In 1527 he printed three *Missals*, one of Sarum use, and in 1528 three Sarum service books and a *New Testament* in Dutch. For 1529 we have no books, and 1530 is represented by an English *Almanack* and two editions of the Sarum *Horae*.

In 1525 Christopher and Hans had printed a *Bible* in Dutch, Hans printing the greater part of the Old Testament and Christopher the remainder and the New Testament. On October 30, 1525, Hans was summoned before the town council of Antwerp for printing a book tainted with Lutheran heresies, was ordered to leave the town and district immediately, and go on pilgrimage to the Holy Blood at Wilsenaken in Prussia, and was further forbidden to return to the town or neighbourhood until he could produce a certificate that the pilgrimage had been carried out. This he apparently objected to do and crossed over to England.

Christopher, left in Antwerp, soon afterwards started on the very dangerous undertaking of printing English New Testaments, which were sent into England and sold there by Hans. In 1528 in the table of certain persons abjured within the diocese of London we find "John Raimund a Dutchman, for causing fifteen hundred of Tyndale's New Testaments to be printed at Antwerp and for bringing five hundred into England." John Raimund is clearly the English form of Jan Roemundt and is probably identical with the Dutchman who earlier in the year was in the Fleet for having sold to

Robert Necton some 200 or 300 copies of the Testament. At the end of 1526 when these Antwerp printed Testaments had found their way into England, a strong effort was made by the English authorities to have them suppressed and the printer punished. On November 24 John Hackett wrote to Wolsey saying there were two printers in Antwerp who printed these books but that a proclamation would soon be issued against them. On the 12th January, 1527, Hackett again wrote that the Margrave had declared that according to the Emperor's last mandment these English books must be condemned to be burnt, the printer Christopher Endhoven banished, and the third part of his goods confiscated. The prisoner's counsel however protested against this judgement, saying that the Emperor's subjects ought not to be judged by the laws of other countries, and his plea seems to have been successful, for Christopher does not appear to have been banished, and continued to print at Antwerp. About the end of 1530 he appears to have crossed over to England in connexion with the sale of English Testaments, and the last act of the drama is tersely stated by Foxe under the year 1531. An Antwerp bookseller named Christopher for selling certain New Testaments in English to John Row, bookbinder, was thrown into prison at Westminster and there died.

On the death of Christopher, Hans wished to return to Antwerp, but in order to do so a certificate of his pilgrimage had to be obtained. In the archives of the city there is an entry dated March 29, 1531, that letters of the pilgrimage having been presented he was free to reenter the city and district. He did not stay long abroad but returned to London, and as an assistant to his fellow-countryman John Nicholson, the printer of Southwark, has been the source of much puzzle to commentators on the English New Testament; for I believe him to be no other than John Hollybush, the so-called reviser of the edition of 1538. The name John Hollybush has generally been considered a pseudonym, and most modern authorities consider the person so designated to have been Miles Coverdale. But Hollybush was a real person, and in the troubled times of 1535, when inhabitants of the Low Countries dwelling in England were forced to become denizened, we find entered amongst others, "John Holibusche alias Holybusche of London, Stationer otherwise bookbinder, born in Ruremund, under the obedience of the Emperor."

Now considering the prominent part that Hans van Ruremond, otherwise John Raimund, had taken in the production and dispersal of the New Testament, he is just the man to be identical with a stationer who revised for the press a new edition. The whole family was identified with the issue of English Testaments. Christopher printed them at Antwerp, and after he had died in England for selling them, his widow continued to print further editions. John had suffered penance for selling them and was now, in more peaceable times, helping to print them. Christopher's son, also named John, became a printer in Antwerp and printed Dutch Bibles and Testaments. One of the very last Sarum service books printed, the *Processional* of 1558, again links Christopher van Endhoven and Christopher van Ruremond together. Upon the title-page is the device with mark and initials, C. E. of Christopher van Endhoven, while the colophon runs "Finit Processionale ad usum Sarum Antverpie impressum per Melchiorem Endovianum, typis Christophori Ruremunden. Anno m.c.lviii. mensis xxiii. Junii."

Another stationer connected with Christopher and Hans van Ruremond was a young man named Peter Kaetz, who acted for a while as their agent in London and had a shop in St Paul's Churchyard. The seven Sarum service books, two *Manuals*, two *Processionals*, a *Horae*, a *Psalter*, and a *Hymni cum notis*, which Christopher van Ruremond printed before February 6, 1525, were all to be sold in London by Kaetz; after this date his place was taken by Birckman. In the spring of 1525 Kaetz returned to Antwerp and carried on business at the sign of the House of Delft in the Cammerstraete, a house formerly occupied by Henri Eckert van Homberg. Here he issued an edition of the *Bible* in Dutch printed by Hans van Ruremond. After this date we hear no more of him. Perhaps like Hans van Ruremond he was involved in the religious troubles, and, banished from Antwerp, found refuge in England. He was in communication with Siberch, the first Cambridge printer, and amongst some fragments rescued from a binding of Siberch's in Westminster Abbey library was a letter addressed to him by Peter Kaetz. In it Kaetz says he is waiting in London for his master, on whose arrival he intends to cross over. He also says, "I send you 25 prognostications and three New Testaments, small size. The prognostications cost one shilling sterling the 25, and the three New Testaments cost 2*s*. and 6*d*. sterling." The puzzling part about the letter is the date to which it should be assigned. Many things point to its being before 1523 and yet the references to the New Testament are not easily explained. The price is that at which the English New Testament printed at Antwerp in 1526 by Christopher van Ruremond was sold, and Hans van Ruremond had come into England about this time to sell it. We know also that Christopher van Ruremond printed prognostications in English. I should have been inclined to put the letter down to some time in or after 1526, and to suppose that the master for whom Kaetz waited was Hans van Ruremond. A person named "Gibkerken" is mentioned in the letter, who may perhaps be a Dutch stationer, John Gybken, who was first an assistant to John Cockes and who afterwards rose to considerable eminence and was admitted as a member of the Stationers' Company. His son was also admitted a member of the Company, although he was deaf and dumb.

The history of the printing and dispersal of the early editions of the English Testament and the various prohibited controversial books offers a very fascinating subject of study, and it is extraordinary, considering the amount that has been written on the question, how little is really known. I suppose the reason is that it has usually been taken up from the theological rather than the bibliographical side, and that writers on the subject though well versed in the history of the Reformation knew nothing of scientific bibliography.

It is obvious that where the printer by his work placed himself in very considerable danger, he would not be anxious to advertise himself and would therefore let his books go out without any imprint or with a fictitious one. Then comes in the study of type. It affords a far more conclusive argument than any theories drawn from where the author was living or where he would probably like to have his book printed. Then again there is the common danger of taking for an early or original edition what is merely a reprint with the original colophon unaltered. An important English library purchased not very long ago at a high price, a so-called first edition of Tindale's *Parable of the Wicked Mammon*, printed at Marlborowe in the land of Hesse by Hans Luft, 1528. But in spite of its having this information emblazoned in gold upon its morocco cover, it is a reprint made about twenty years later by Berthelet, a fact that should have been patent to anyone at all conversant with English printing, seeing that the title-page is within one of Berthelet's best-known border frames.

Then, again, many of the books with foreign imprints were certainly printed in this country. Upright Hoff of Leipsic and Ian Troost of Aurich are both names assumed by John Oswen, the printer of Ipswich.

Cambridge has made an important start towards a real bibliography of this subject by the publication of the third volume of Mr Sayle's most admirable *Catalogue of the Early English Books in the University Library*. Only those who have worked on the subject can understand how much patience, how much untiring labour, and how many often disappointing searches must have been expended on every small fact chronicled in the volume.

To return to these Early English Testaments printed

in Antwerp. It is quite clear that as soon as the edition printed by Schoeffer in 1525 had been issued and found a ready sale, the Antwerp printers set to work to print rival editions. Fortunately we have a very clear and contemporary account which there seems no reason for doubting. It occurs in Joye's *Apology*, written about 1534, and I will abbreviate it as far as possible. "Thou shalt know that Tyndale translated and printed the New Testament in a mean great volume but yet without Kalendar, Concordances in the margin, and Table in the end. And anon, after, the Dutchmen got a copy and printed it again in a small volume adding the Kalendar in the beginning, Concordances in the margin, and the Table in the end." After speaking of the many faults in it he continues: "After this they printed it again also without a corrector in a greater letter and volume with the figures in the Apocalypse which was therefore much falser than their first. When these two prints (there were of them both about five thousand printed) were all sold more than a twelvemonth ago [that is about the middle of 1533] Tyndale was pricked forth to take the Testament in hand. But Tyndale prolonged and deferred in so much that in the mean season the Dutchmen printed it again the third time in a small volume like their first print, but much more false than ever it was before." Before this edition was put in hand Joye had been asked to correct it, but refused, saying that Tindale was about to bring out a revised edition: however two thousand copies were printed and sold almost at once. A fourth edition was then put in hand and Joye was again asked to correct it. He continues: "I answered as before that if Tyndale amend it, with so great diligence as he promiseth yours will be never sold. Yes, quod they, for if he print two thousand and we as many, what is so little a number for all England? and we will sell ours better cheap and therefore we doubt not of the sale." Joye seeing clearly that they were determined to print whether he corrected or not, and seeing he was more qualified than a native, undertook the task, and for the modest remuneration of fourpence halfpenny for every sheet of sixteen leaves corrected the edition published by the widow of Christopher van Endhoven in August, 1534. According to this statement, which is a very lucid one, between the issue of Tindale's octavo version in 1525 and Joye's revision in 1534, three other Antwerp editions had been issued. The first in a very small volume with a Kalendar, marginal notes, and a table. This must be the edition printed by Christopher van Endhoven in 1526 for which he was prosecuted. The second was of a larger size and had woodcuts to the Apocalypse. The third was similar to the first, but much more incorrectly printed. Here then are three editions as yet entirely unidentified, and there are most probably more. The late Dr Angus possessed the title-page of an edition of 1532 which from the tracing reproduced in Demaus' book on Tindale must have been a quarto.

Antwerp continued to print Testaments in increasing numbers. In 1534 and 1535 four editions were issued, and in 1536 when they were more freely circulated, no less than seven. These are beyond our period, but I cannot resist chronicling a fact lately come to light, which gives the solution of a long-standing puzzle. In all the three quarto editions of 1536, none of which has a printer's name, is a woodcut of St Paul with his foot resting on a stone. These cuts though almost exactly similar are not identical. In one cut nothing is engraved on the stone, and the edition containing it is known as the "blank stone" edition. In another cut a mole is engraved on the stone, and the edition is known as the "mole" edition. In the third there is what is called an engraver's mark, with the initials A. K. B., and that issue is called the "engraver's mark" edition. Now about this engraver's mark nothing was known. Not long ago, however, in turning over a large volume of miscellaneous woodcuts in the Bodleian I came across an example of this cut on what was manifestly the last leaf of a book. Below the cut was this colophon, "Antwerp. Mattheus Cromme voor Adriaen Kempe van Bouckhout, 1537." Between 1537 and 1539 Crom printed for Kempe several editions of Branteghem's *Vita Jesu Christi*. In 1536 the cut occurs in the *Storys and prophesis out of the holy scriptur* printed by Simon Cock. It would thus seem that Kempe first employed Cock as his printer, but left him for Crom when the latter began printing about 1537.

The specimens of bookbinding of this period, the best so far as stamped bindings are concerned, offer a very difficult problem to the student. What are English bindings and what are not, and how are they to be distinguished? The more the subject is studied, the more difficult these questions appear. There are bindings which are obviously English just as there are others obviously foreign, but the great majority are doubtful. It must be remembered that a very large number of binders in England were foreigners, who would very likely bring their dies and stamps with them from abroad, and finish their work in a foreign manner. As their bindings were produced in England, they must be called English bindings though in a foreign style, for we would not call the Berthelet bindings, produced for Henry VIII in London by foreign workmen with foreign dies, anything but English.

As an example of this confusion I would mention the binding on a copy of the *Horae ad usum Sarum* of 1506, printed for Pelgrim and Jacobi, and now in the Bodleian. In the centre of the cover are two purely Netherlandish panels, while round these runs a frame composed of two dies which had belonged to Caxton. We know that the partners owned several of Caxton's dies, because they are found in conjunction with Jacobi's signed panels. As the *Horae* was printed for them and has their dies on the binding, it is clear they are the binders, but had the Caxton dies been absent, no one could have put down the binding as anything but Netherlandish.

The favourite English binding of this time consisted of two panel stamps, one having the Royal Arms, the other the Tudor Rose with mottoes, which I described before. In the base of these panels the binder's initials and mark generally occur. In some cases these initials can with certainty be ascribed, because they occur with the mark and full name in some printed book. Thus there is no mistake about the bindings of Pynson, Notary, Jacobi, or Reynes. But in a large number of cases there are merely initials, or initials with a mark which has not been identified. Here the imagination steps in with usually fatal results.

One set of these panels is signed R. O., and as O is an uncommon initial for a surname the natural inclination is to attribute the binding to Reginald Oliver. Another set has the initials R. L.: who more likely than Richard Lant ? All such suggestions are worthless without proof, and until proved are not worth setting down. Other binders using these panels were H. N. and G. G.

A binder whose initials were G. R. had two very handsome pairs of panels. One panel contains the Royal Arms surrounded by the verse beginning, "Confitemini dominum quoniam bonus," while the corresponding panel contains figures of four saints. The first of the second pair is also the Royal Arms with the verse beginning "Laudate Dominum de terra," and the other has the figures of the four saints, George, Barbara, Michael, and Katherine.

The Annunciation was a very favourite subject with bookbinders. One of the handsomest panels of this kind was used by a binder, A. H., who used also a panel with the Tudor Rose. Though this shows that the bindings were produced for the English market, the leather employed, apparently sheepskin, suggests that they were executed abroad, perhaps at Rouen.

Two other panels were in common use containing figures of St Sebastian and St Roche. These were extremely popular, for at that time the country was periodically visited with outbreaks of the plague, against which these two saints were considered the guardians. Very often the saint chosen for the binder's panel had reference to his own Christian name. John Reynes has a panel with St John Baptist. St George as the patron saint of England was naturally popular. One beautiful binding has on one side St George and the Dragon, and on the other St Michael with the binder's devices, a maiden's head on a shield. This might be the work of Richard Faques, who lived at the sign of the Maiden's Head.

The roll binding, as apart from the panel binding, was ornamented by means of a rolling tool, and at first these were large and ornamental, but they gradually narrowed, and so became meagre or weak in detail.

Those in use at Cambridge are perhaps the finest of all, but this is not the time to speak of them. Among London binders John Reynes produced the best known roll, containing figures of a hound, a falcon, and a bee with flowers and foliage. Another with an unknown binder's mark, contains a hound, a falcon, and a double-headed eagle on a shield. The roll of another unknown binder bears a Tudor Rose, a pomegranate, a shield with the arms of France and England, a turreted gateway, a portcullis, a fleur-de-lys, and the binder's mark.

Another handsome roll has the initials of I. G. and W. G. with figures of a dragon, a gryphon, and foliage, and is not at all uncommon. In Archbishop Marsh's library at Dublin there is an example of such a binding, tooled entirely in gold, the only example I have ever seen of such work.

In the main design of all these English rolls there is very little variation, the various royal emblems amidst foliage and flowers.

It is unfortunate that while we have so many initialled bindings which cannot be definitely ascribed to particular binders, so also we have the names of plenty of binders to whom we cannot allot bindings; Alard and Noel Havy, De Worde's binders, Giles Lauret, who lived close to Havy in Shoe Lane, and John Rouse, who lived next door. John Richardson, and John Row who was punished for buying Testaments. John Pollard and Thomas Stoke who received pardon for some unknown offence in 1533. Martin Dotier, the third offender, we do find afterwards, for he commissioned one edition of a Sarum *Manual* in 1543, and ended his days as a brother of the Stationers' Company.

Few binders or stationers of the time have left any definite record. I have notes of at least four hundred persons connected with the English book-trade between the years 1500 and 1535, but of very few is anything known beyond the name. A mention in a will, an entry in a record or tax-roll, or a mention in church or municipal accounts, is all that remains to tell us of their existence.

I have referred before to the various difficulties with which the foreign stationer had to contend, and year by year these difficulties grew worse. The educated man was still entirely dependent on the Continent for such books as he required, for before 1535 the literature of the Renaissance was untouched by the native printers. Wynkyn de Worde from start to finish, roughly speaking, never printed a classic, but was content to turn out rhymes and romances to catch the popular taste. Pynson, the more scholarly worker, was engaged on official or semi-official publications. There was no press that could print in any but ordinary type; even the most learned printer had not sufficient Greek letters to print quotations; and this when foreigners came to England to learn Greek.

No doubt the large importation of books was galling to the native printers, but they do not seem to have lacked work, or if they did it was only through their own want of energy. The bookbinders were in a much worse case, because so many binders in this country were foreigners, and no doubt large consignments of books came over ready bound. The probability is that the foreigners had been better trained, and could turn out cheaper and better work, and though that might appeal to the purchaser, it would certainly injure their native rivals. While for forty years the Italians had been issuing beautiful little volumes in gilt bindings, the use of gilding on bindings was almost unknown in England. It was not as though English printing was improving as time went on. Indeed it became more and more careless and slovenly. A printer here and there who had a subsidy and was sure of work, such as Pynson and Berthelet, cast fine type and produced handsome books. But the majority lagged far behind, and worn-out type and broken cuts were the stock-in-trade of the ordinary printer. Then was passed the Act of 1534, which declared the English printers to be at least the equals of any foreign competitors, and restricted the importation of foreign books. With the removal of the foreign competition native work sank to its lowest level, and it was only when the religious persecutions abroad drove numbers of refugee foreign printers to England that English printing began to revive—for the time.

On Christmas Day, 1534, probably but a few days

before the death of Wynkyn de Worde, the famous Act of 25 Henry VIII concerning printers and binders of books came into operation. The preamble is extremely clear and interesting: "Whereas by the provision of a statute made in the first year of the reign of king Richard III [1484] it was provided in the same Act, that all strangers repairing into this realm, might lawfully bring into the said realm printed and written books to sell at their liberty and pleasure. By force of which provision there hath come into this realm sithen the making of the same a marvellous number of printed books, and daily doth; and the cause of making of the same provision seemeth to be, for that there were but few books, and few printers, within this realm at that time, which could well exercise and occupy the said science and craft of printing; nevertheless, sithen the making of the said provision, many of this realm, being the king's natural subjects, have given themselves so diligently to learn and exercise the said craft of printing, that at this day there be within this realm a great number of cunning and expert in the said science or craft of printing, as able to exercise the said craft in all points, as any stranger in any other realm or country. And furthermore, where there be a great number of the king's subjects within this realm, which live by the craft and mystery of binding of books, and that there be a great multitude well expert in the same, yet all this notwithstanding there are divers persons, that bring from beyond the sea great plenty of printed books, not only in the Latin tongue, but also in our maternal English tongue, some bound in boards, some in leather, and some in parchment, and them sell by retail, whereby many of the king's subjects, being binders of books, and having no other faculty, wherewith to get their living, be destitute of work, and like to be undone, except some reformation be herein had. Be it therefore enacted by the king our sovereign lord, the lords spiritual and temporal, and the commons in this present parliament assembled, and by authority of the same, that the said proviso, made the first year of the said king Richard III, from the feast of the nativity of our Lord God next coming, shall be void and of none effect."

This part of the Act, which seems to me essentially fair, simply removed special privileges and put the foreign printers and stationers on an equal footing with all other aliens. Then come two special new enactments.

"And further, be it enacted by the authority aforesaid, that no persons resiant or inhabitant within this realm, after the said feast of Christmas next coming, shall buy to sell again, any printed books, brought from any parts out of the king's obeysance, ready bound in boards, leather or parchment, upon pain to lose and forfeit for every book bound out of the said king's obeysance, and brought into this realm and bought by any person or persons within the same to sell again contrary to this Act, 6s. 8d."

This enactment was a great protection to native binders as prohibiting all dealing in foreign bound books. The next paragraph was for the benefit of the printers. "And be it further enacted, by the authority aforesaid, that no person or persons inhabitant or resiant within this realm after the said feast of Christmas shall buy within this realm of any stranger born out of the king's obedience, other than of denizens, any manner of printed books brought from any the parts beyond the sea, except only by engross and not by retail upon pain of forfeiture of 6s. 8d. for every book so bought by retail contrary to the form and effect of this estature."

The remainder of the Act treats only of penalties and restrictions as to the price of books and bindings.

The clause about printing amounts to this, that no person not a native or denizen could retail foreign printed books, and it seems pretty clear that it was not so much intended for the advantage of printers, as for giving fuller power to suppress the importation and surreptitious sale of controversial books.

The alien printer, stationer, or binder, when denizened, was in this position. He paid double subsidies and taxes, he could have none but English-born apprentices and only two foreign workmen. He was under the rule of the Warden of the Craft. He could not deal in foreign bound books, nor buy books from foreigners except by engross. The alien not denizened had the further restrictions that unless he had been a householder before February 1528, he could not keep any house or shop in which to exercise any handicraft, nor could he sell any foreign printed books by retail.

Weighed down by these various disabilities the small foreign stationers rapidly disappeared. A certain number who had lived here for some time took out letters of denization and stayed on, but it was only to struggle on.

If you wish to have a practical idea of what foreign work meant to England, just glance at the list of Contents in Mr Sayle's admirable *Catalogue of the Early*

English books in the University Library. Taking all the persons given in the list, from the earliest printing to 1510 connected with English books, we have the astonishing result of three Englishmen, Caxton, a printer, Hunte, the Oxford stationer, and W. Bretton, merely a patron of printing. The foreigners on the other hand number forty-two. After 1510 the relative numbers rapidly alter, and by 1535 the great majority were English. The hatred of the natives, and the various Acts levelled against him, compelled the foreign workman to disappear, and I am afraid that with his disappearance the necessary competition which produced good printing disappeared also.

The fifty years of freedom from 1484 to 1534 not only brought us the finest specimens of printing we possess, but compelled the native workman, in self-protection, to learn, and when competition was done away with his ambition rapidly died also. Once our English printing was protected, it sank to a level of badness which has lasted, with the exception of a few brilliant experiments, almost down to our own day.

INDEX.

Aberdeen University Library, 81
Abridgement of the Statutes, 44, 46, 114, 124
Actors, Peter, 77, 98, 169
Acts affecting printers and stationers, 74, 129, 187, 189-191, 208, 212, 236-239
Adryan of Barrowe. *See* Berghen, A. van
Advertisement, Caxton's, 9
Aeneas Sylvius. See *Eurialus*
Aesop, *Fables*, Caxton, 15; Andrewes, 156
Alard, bookbinder, 107, 139, 140, 235
Albertus, *De modis significandi*, Notary, 37; *Liber aggregationis*, Machlinia, 48; *Secreta mulierum*, Machlinia, 48
Alcock, John, *Gallicantus*, Pynson, 1498, 66; *Mons perfectionis*, Pynson, 1497, 66; W. de Worde, 1501, 132
Aldus bindings, 124
Alexander VI., pope, 55
Alexander Grammaticus, *Doctrinale*, Pynson, 59, 60, 66, 109
Alien printers and stationers, 74-76, 120, 188-191, 235-240
Almanack, 1529, A. van Berghen, 216; 1530, C. van Ruremond, 223
Alost, 5
Althorp Library, 7, 8, 9, 10, 20, 114

Ames, Joseph, 8, 11, 50, 152, 171
Andreae, A., *Questiones*, Lettou, 1480, 42
Andrewe, Lawrence, 140, 155, 156, 221
Anelida and Arcyte, Caxton, 10
Angus, Dr, 230
Antiphoner, Sarum, W. Hopyl, 205, 218
Antoine, Jean, 206
Antwerp printing, 79, 85, 88, 91, 148, 149, 153, 155, 176, 200, 213
Antwerp stationers, 214-231
Anwykyll, J., *Grammar*, 79
Appleby Grammar School, 60
Arnold, Richard, 221; *Chronicle*, 169, 214-216
Ars Moriendi, Caxton, 21, 22, 90
Art and Craft to Know Well to Die, Caxton, 64; Pynson, 64
Arundel, Earl of, 14, 16
Ashburnham sale, 30, 53
Atkyns, R., *Original and growth of printing*, 1

B., A. K., engraver, 231
Badius, J., 96, 193, 195
Bagford, John, 8, 11, 148
Baldwin, John, 193
Baligault, Felix, 84, 86
Bankes, Richard, 125, 154, 155
Barbanson, John, 139
Barbier, Jean, 37, 38, 194, 196

Barclay, Alexander, 162, 206;
  Introductory to French, 147
Barker, Christopher, 76
Barnes, Robert, 125
Barrevelt, Gerardus, 92, 93
Bars, John, 69
Bartholomæus, *De proprietatibus rerum*, 4, 29
Barton, Elizabeth, 151
Basle, 78
Bath, Marquis of, 54
Battle of Agincourt, J. Skot, 151
Beauchamp, Earl, 66
Bedford, Duke of, *Epitaph of*, 63
Bedfordshire General Library, 106
Bellaert, Jacop, 88, 89
Berclaeus, Thomas, 177
Bercula, Thomas, 177
Berghen, Adrian van, 91, 214–216
Bernard, St, *Meditations*, W. de Worde, 90
Berthelet, Anthony, 183
Berthelet, Edward, 183
Berthelet, Margaret, 182
Berthelet, Thomas, 156, 157, 158, 175, 176, 177–183, 192, 208, 228, 236; bindings, 124, 232
Bible, Grafton and Whitchurch, 1539, 209; Petyt and Redman, 1540, 176; Dutch, C. & H. van Ruremond, 1525, 223, 226
Bible, New Testament, English, 208, 222–231; Dutch, 1528, 223
Bibliothèque Nationale, 68, 73, 136, 137
Binder, royal, 181
Binders, 97, 101–126, 231–235
Bindings, 101–126, 231–235
  Armorial, 123
  Cambridge, 108, 117, 120, 122, 234
  Characteristics, 122–3
  Clasps and Ties, 123
  Durham, 102, 104
  Early English, 101–111
  Foreign, importation prohibited, 76, 238
  Fragments in, 107, 110, 124–5

Bindings (*continued*):
  Gilt, 182, 234, 236
  Greek books, 123
  Italian, 123
  London, 102
  Norwich, 108
  Oxford, 103–4
  Panel stamps, 111–121, 145, 197–8, 200–1, 232–4
  Prices, 182
  Roll ornaments, 120, 200, 202, 234
  Rouen, 121, 233
  Venetian style, 124, 182
  Winchester, 102–4
Birckman, Arnold, 198, 200, 201, 219
Birckman, Francis, 195, 217–219, 222, 226
Birckman, Francis, jun., 218, 219
Birckman, John, 201, 219
Birth of Merlin, W. de Worde, 1510, 136
Blades, Wm, 3, 8, 19, 22, 23, 58, 106, 107
Blanchardin and Eglantine, Caxton, 19, 20
Blandford, Marquis of, 9, 61
Bocard, André, 97, 142, 193
Bodleian Library. *See* Oxford
Body of Policy, J. Skot, 1521, 150
Boethius, A. M. T., *De consolatione philosophie*, Caxton, 106–7
Bonaventura, *Speculum vitae Christi*, Caxton, 16, 20, 58; W. de Worde, 26; Pynson, 58, 62
Bonham, William, 149, 184
Book of Cookery, Pynson, 1500, 69, 71, 159; Bankes and Lant, 1545, 155
Book of Courtesy, Caxton, 10, 25; W. de Worde, 24, 25
Book of Divers Ghostly Matters, Caxton, 21
Book of Dystyllacyon, 155, 221
Book of Herbs, J. Skot, 151
Book of Justices of Peas, R. Redman, 176

INDEX

Book of Kerving, W. de Worde, 134
Book of maid Emlyn, J. Skot, 150
Book of St Alban's, W. de Worde, 1496, 29
Book of Songs, W. de Worde, 1530, 138
Book to lerne to speak French, Pynson, 65
Book trade in sixteenth century, 187–192
Bookbinders. See Binders
Borbonius, Nicholas, 181
Borde, Andrew, 147
Borders, 21, 27, 38, 49, 56, 157, 169, 176, 180, 194
Boteler, John, 152
Boudins, John, 97, 193
Bourman, Nicolas, 140
Brachius, John, 206
Bradshaw, Henry, 8, 11, 12, 18, 19, 21, 43, 50, 80, 99, 108, 109, 125, 156
Brand sale, 9
Brant, S., *Ship of Fools*, Pynson, 162; W. de Worde, 135
Branteghem, G. de, *Vita Jesu Christi*, 231
Braunschweig, H. See Jerome of Brunswick
Bretton, William, 194, 195, 196, 204
Breviary:
  Aberdeen, 1509–10, 7
  Hereford, 1505, 82
  Sarum, 92; Cologne, ab. 1475, 73; Venice, 1483, 73; Paris, 1494, 84; Venice, 1495, 93; Rouen, 1496, 81; Louvain, 1499, 80; Pynson, 1507, 162; 1507, 206; Antwerp, 1526, 223; Paris, 1531, 205
  York, 1493, 92
Bright, B. H., 61
British Museum, 8, 9, 14, 16, 20, 44, 49, 50, 53, 54, 58, 60, 61, 68, 69, 70, 81, 95, 105, 109, 114, 134, 138, 142, 159, 165, 170, 172, 184, 186, 196, 217

Brito, John, 52
Britwell Library, 40, 145
Bull of Innocent VIII., Machlinia, 55
Bulle, J., 41
Bumgart, Herrman, 115
Burgo, J. de, *Pupilla Oculi*, Paris, 1510, 194
Burgundy, Duchess of, 3
Butler, John, 139, 152
Byddell, John, 138, 139, 140, 203, 204

C., N., binder's stamp, 116
Caesar, *Commentaries*, W. Rastell, 186
Caillard, J., 205
Cambridge bindings, 108, 117, 120, 122, 216, 234
Cambridge printing, 11, 99, 125
Cambridge libraries:
  Caius College, 54, 86, 115
  Christ's College, 165
  Clare College, 156
  Corpus Christi College, 11, 103, 110, 161, 170, 212, 216
  Emmanuel College, 161, 170
  Jesus College, 43, 108
  King's College, 62
  Magdalene College, 186
  Pepysian Library, 14, 28, 34, 61, 63, 149, 163
  St Catharine's College, 81
  St John's College, 92, 110, 209
  Trinity College, 68, 96
  University Library, 4, 8, 10, 25, 32, 35, 36, 37, 38, 44, 48, 49, 51, 53, 61, 73, 80, 82, 84, 88, 92, 95, 96, 116, 132, 143, 147, 150, 153, 165, 172, 206
Campensis, Joannes, *Interpretatio Psalmorum Omnium*, 1534, 208
Campion, Amye, 166
Campion, Joane, 166
Campion, William, 166
Candos, Guillaume, 206
Canutus, *Treatise of the Pestilence*, Machlinia, 53

Caorsin, Gul., *Siege of Rhodes*, 45
Capgrave, J., *Nova Legenda Angliae*, W. de Worde, 137
Carmelianus, Petrus, 79, 164
Castellain, George, 69, 161
Castle of Labour, A. Verard, 206
Castle of Pleasure, H. Pepwell, 1518, 147
Cato, Caxton, 10, 12
Caxton, Elizabeth, 23
Caxton, Maud, 20
Caxton, W., 3-24, 46, 57, 58, 73, 78, 88, 90, 130, 146
  Binding stamps, 108, 139, 140, 197, 232
  Bindings, 105-107
  Border used by, 21, 27
  Death, 23
  Device, 18, 38, 58
  Family, 23
  Foreign reprints of his books, 24
  Number of books printed, 22
  Situation of printing office, 6, 39
  Translations by, 14, 23, 28
  Types, 6, 12, 22, 26, 56
  Vellum books, 20, 125
  Wife, 20
  Woodcuts, 12, 13, 15, 16, 17, 21, 22, 27, 33, 142
Charles the Great, Caxton, 16
Chasteleyn, George. *See* Castellain, George
Chastising of God's Children, W. de Worde, 24, 25
Chatsworth Library, 81, 88
Chaucer, G., *Works*, Pynson, 1526, 165; Godfray, 1532, 156, 157; *Canterbury Tales*, Caxton, 8, 15; Pynson, 57, 58; W. de Worde, 30; *Hous of Fame*, Caxton, 15; *Mars and Venus*, Notary, 39; *Troilus and Cressida*, Caxton, 15
Chepman, Walter, 115
Chevallon, Claude, 205
Cholmondeley, Ralph, 176
Chorle and the Bird, 10, 16

Christmas Carolles, W. de Worde, 1521, 137
Chronicles of England, Caxton, 13; Leeu, 88, 90; Machlinia, 52, 60; Notary, 142-144
Cicero, *Paradoxes*, Redman, 177; *De officiis*, Mainz, 1466, 4; *Pro Milone*, Oxford, 66
Claudin, A., 17
Cluen, Gerard, 217
Coblentz, Jean de, 198
Cock, Simon, 231
Cockes, John, 227
Colet, John, 148
Cologne printing, 4, 65, 73, 79, 142, 219
Combe, Dr Charles, 73
Commemoratio lamentationis beate Marie, Caxton, 21
Commendations of Matrimony, J. Skot, 1528, 150
Complaint of the too soon maryed, W. de Worde, 138
Confluentinus, Joannes, 198
Congregational Library, London, 61
Consolation of timorouse and fearfull consciencys, 172
Constable, John, *Epigrammata*, Pynson, 1520, 124
Contemplacyon or meditacyon of the shedynge of the blood, W. de Worde, 35
Conversion of Swearers, W. de Worde, 135; J. Butler, 152
Conway, Sir W. M., 88
Copenhagen printing, 30
Copland, Robert, 7, 139, 146-7, 154, 172
Copland, William, 147
Corsellis, Frederick, 2
Cotton, Henry, 19
Cousin, Jacques, 205, 206
Couvelance, Philippus de, 199
Coverdale, Miles, 208, 209, 225
Cowlunce, Jean de, 198
Cox, Leonard, 153
Cranmer, Thomas, 159
Crawford, Earl of, 194
Criblée engravings, 142

INDEX

Crom, Matthew, 231
Cromwell, Thomas, 149, 154, 157, 185, 203, 209
Croppe, Gerard, 23
Cuthbert, St, 101

Darby, Robert, 139
Dating, method of: Berthelet, 179; Notary, 135, 141; Pynson, 68, 159, 160; Redman, 175; W. de Worde, 31, 135
Davidson, Thomas, 115
Day, John, 191
De veteri et novo Deo, J. Byddell, 1535, 203
Debate and stryfe betwene Somer and Wynter, L. Andrewe, 156
Defence of Peace, 1535, 203
Demaundes Joyous, W. de Worde, 1511, 136
Determinations of the most famous Universities, Berthelet, 179
Deventer printing, 79
Devonshire, Duke of, 16, 74. *See also* Chatsworth Library
Dewes, G., *Introductorie for to lerne French*, T. Berthelet, 157; J. Reynes, 200
Dialogue betwixte two englyshe men, T. Berthelet, 180
Dibdin, T. F., 8, 27, 46, 53
Dictes or sayengis of the Philosophres, Caxton, 6, 7
Dictionary of National Biography, 22, 131
Directorium sacerdotum. *See* Maydeston, C.
Directory of the conscience, L. Andrewe, 156
Diurnale, Sarum, W. Hopyl, 1512, 195
Dives and Pauper, Pynson, 54, 61
Division of the Spiritualty and the Temporalty, 175, 180
Dockwray, Thomas, 148, 149
Doctrinal of Sapience, Caxton, 20
Doctrynale of good servantes, J. Butler, 152

Doesborch, Jan van, 91, 130, 155, 214, 220–222
Donate and accidence, Paris, 1515, 198
Donatus, P. Violette, 206
Donatus Melior, Caxton, 17, 125
Dorne, John, 98
Dorp, R. van den, 220
Dotier, Martin, 235
Douce, Francis, 10, 25, 221
Douglas, Gavin, 20
Draper, Richard, 202
Drunkardes, The IX, R. Bankes, 154
Dublin: Marsh Library, 143, 234; Trinity College Library, 84, 86, 88, 89, 217
Duff, E. Gordon, 60, 115, 116
Du Pré, Jean, 206
Durham bindings, 102, 104
Durham Cathedral Library, 34, 102
Dying Creature, W. de Worde, 1514, 146

Eckert van Homberg, Henri, 226
Eckius, J., *Enchiridion*, 1531, 148
Edinburgh printing, 151
Edinburgh, Advocates' Library, 37
 Signet Library, 67
 University Library, 81
Edwards, bookseller, 161
Egmond, Count of, 67
Egmont, Frederick, 66, 67, 91–94, 97, 207; bindings, 114–5
Elegantiarum viginti praecepta, Pynson, 67
Elyot, Sir T., *Book named the Governour*, T. Berthelet, 180
Endhoven, C. van. *See* Ruremond, C. van
Eneydos, Caxton, 20
Epitaph of Jasper, Duke of Bedford, Pynson, 63
Erasmus, D., *Christiani hominis institutum*, H. Pepwell, 148; *Colloquiorum formulae, De copia verborum, Enchiridion militis christiani*, W. de Worde,

246 INDEX

Erasmus, D. (*continued*):
138; *Exposition of the commune crede*, Redman, 203; *Good manners for children*, W. de Worde, 138; *Treatise upon the pater noster*, Berthelet, 178
Esteney, John, 130
Eurialus and Lucrece, J. van Doesborch, 220
Every Man, J. Skot, 150
Exposicions des epistres et evangiles, Verard, 1511-2, 212
Expositio hymnorum, A. Bocard for J. Boudins, 97, 193; H. Quentell, 65; Pynson's Supplement, 65
Expositiones terminorum legum Anglorum, 1527, 152

Faques, Richard, 170-172, 234
Faques, Wm, 158, 162, 169-171
Far, Richard, 172
Farmer, Richard, 10, 39, 132
Fawkes, Michael, 172
Fawkes. *See also* Faques
Faxe, Amelyne, 172
Faxe, Richard, 172
Ferreboue, James, 212
Festum nominis Jesu, Pynson, 61, 65
Festum transfigurationis, Caxton, 61; Machlinia, 54; Pynson, 65
Festum visitationis, Machlinia, 54
Fewterer, J., *Myrrour of Christes Passion*, R. Redman, 175
Ficinus, M., *Epistolae*, 1495, 103
Fifteen Joys of Marriage, W. de Worde, 135
Fifteen Oes, Caxton, 21, 22, 27
Fifteen Tokens, J. van Doesborch, 220
Fisher, John, *Sermon*, W. de Worde, 1508, 134
Fitzherbert, Sir A., *Diversite de courtz*. R. Redman, 1523, 172; *Great Abridgement*, J. Rastell, 184
Fitzjames, R., *Sermo die lune*, W. de Worde, 28
Fletewode sale, 35

Foreign book-trade with England, 72-100, 187-8, 205-213, 214-231, 235-240
Foundation of Our Lady's Chapel at Walsingham, Pynson, 63, 64
Four Sons of Aymon, Caxton, 19, 20
Frankenberg, Henry, 77
Frankfurt fair, 192
Frederyke of Jennen, J. van Doesborch, 220, 221
Freeling, Sir F., 61
Froissart, J., *Chronicle*, Pynson, 164
Froschover, Christopher, 218
Frute of Redempcion, R. Redman, 175
Fryth, John, *Disputation of Purgatory*, 184

G., E., engraver, 172
G., G., bookbinder, 233
G., I., bookbinder, 234
G., W., bookbinder, 49, 109, 234
Gachet, John, 212
Galfridus Anglicus, 79
Game and Playe of the Chesse, Caxton, 6, 12
Garlandia, J. de, 63, 79
Gaver, James, 107, 139-141
Gavere, Ioris de, 112
Ghent binding, 112
Ghent University Library, 21
Gibkerken, 227
Gift of Constantine, T. Godfray, 157, 203
Gloucester Cathedral Library, 82
Godfray, Thomas, 156, 157, 203
Golden Legend. *See* Voragine, J. de
Golden Litany, J. Skot, 151
Gottingen University Library, 9
Gouda printing, 30
Gough, John, 139, 184, 203, 204
Gough, Richard, 92, 199
Gourmont, Egidius, 196
Governayle of Health, Caxton, 90
Gower, J., *Confessio amantis*, Caxton, 15
Gradual, Sarum, 1527, 199, 205

Gradus comparationum, J. Toy, 1531, 150, 153
Graf, Urs, woodcuts by, 211
Grafton, Richard, 155, 181, 208, 209
Gray, William, 154, 155
Greek type, 235
Grenville Library, 61
Gringore, P., *Castle of Labour*, Verard, 206
Growte, John, 204
Groyat, John, 204
Gryphus, P., *Oratio*, Pynson, 163
Gueldres, Duke of, 67
Guilford, Sir Richard, 163
Guilibert, John, 112
Gulielmus de Saliceto, *Salus corporis salus anime*, R. Faques, 171
Guy of Warwick (Pynson), 70
Gybken, John, 227

H., A., bookbinder, 121, 233
H., I., printer, 37, 38
Hackett, John, 224
Haghe, Ingelbert, 82
Hain, L., *Repertorium Bibliographicum*, 39
Halberstadt Library, 14
Hampole, Richard de, *Devout Meditacions*, 134; *Speculum Spiritualium*, 194
Hardouyn, Gilles, 205
Haukins, John, 158, 166, 167, 168
Havy, Noël, 139, 140, 235
Hawes, S., *Pastime of Pleasure*, W. de Worde, 1509, 135
Hazlitt, W. C., 136
Heber sale, 35, 40
Heerstraten, E. vander, 77
Helias, Knight of the Swan, W. de Worde, 1512, 136
Henry VII., 55, 68, 212
Henry VIII., 68, 164, 165, 212
Herbal, *The Grete Herball*, 1529, 156
Herbert, William, 35, 39, 114, 143, 152, 169, 174, 178, 204, 207
Hereford bookseller, 82, 83
Herford, John, 149

Herolt, John, *Sermones discipuli*, J. Notary, 1510, 143
Heron, John, 184
Hertzog de Landoia, Joh., 91–93
Heywood, J., *Gentleness and Nobility*, J. Rastell, 185; *Johan the Husband, Pardoner and the Friar, Play of Love, Play of the Weather*, W. Rastell, 186
Hieronymus de Sancto Marcho, *De universali mundi machina*, Pynson, 161
Higden, R., *Polycronicon*, Caxton, 13; Treveris for Reynes, 1527, 199
Higman, J., 18, 205
Higman and Hopyl, 87
Hillenius, Michael, 148, 176
Hilton, W., *Scala perfectionis*, J. Notary, 1508, 143
History of Jacob, J. Skot, 150
Hoe, Robert, 16, 136
Hoff, Upright, 228
Holder, Robert, 201
Holkham Library, 26
Hollybush, John, 225
Holt, J., *Lac Puerorum*, A. van Berghen, 91, 216; J. van Doesborch, 220
Holwarde, Thomas, 201
Homiliarius (? Cologne, ab. 1475), 73
Hopyl, Wolfgang, 84, 87, 95, 96, 194–196, 205, 218
Horae, Paris editions, 84–86; undated editions, 85; J. Poitevin, 86
Horae, Sarum: number of editions, 85; Caxton, 17, 21, 33; Leeu, 80; Machlinia, 48, 49, 109; Notary, 38, 39; C. van Ruremond, 226; W. de Worde, 27; Venice, 1494, 91; Paris, 1498, 96; 1506, 232; 1507, 194; Paris, 1510, 194; Paris, 1532, 1533, 1534, 204; Rouen, 1536, 204; Antwerp, 1530, 223
Horologium Devotionis, Zel, 142
Horse the Shepe and the Goose, Caxton, 10; W. de Worde, 22

Howleglas, 89; J. van Doesborch, 220
Hundred mery tales, J. Rastell, 184
Hunte, Thomas, 98
Hunterian Museum, Glasgow, 19, 64, 155
Huvin, Jean, 37, 38
Hylton, W., *Scala perfeccionis*, W. de Worde, 26
Hymni cum notis, C. van Ruremond, 226
Hymns and sequences, J. Notary, 143

Imitatio Christi, Pynson, 114, 160
Imposition, wrong, instance of, 50
Indulgences, 104, 106; Caxton, 12, 19; Lettou, 12, 43, 108
Infancia Salvatoris, Caxton, 9
Informatio Puerorum, Pynson, 69
Information for Pilgrims, W. de Worde, 28
Initial letters, 93, 142; filled in by hand, 51
Inner Temple Library, 39
Innocent VIII., 55
Institution of a Christian Man, T. Berthelet, 1537, 180
Interlude of the four elements, J. Rastell, 185
Interlude of women, J. Rastell, 185
Introductorium linguae latinae, W. de Worde, 28
Ipswich, 228

Jacobi, Henry, 105, 108, 148, 194–199, 232; bindings, 119, 197, 198
Jacobus, illuminator, 112
Jean le Bourgeois, 169
Jeaste of Sir Gawayne, J. Butler, 152
Jehannot, Jean, 96
Jerome of Brunswick, *Boke of Distillacyon*, Andrewe, 155, 221
Joannes de Lorraine, 82

John of Aix-la-Chapelle, 98
John Rylands Library, 26, 30, 53, 55, 68, 84, 161, 162. See also Althorp Library
Johnson, Maurice, 152
Joye, G., 229, 230
Justice of Peace, R. Copland, 1515, 147

Kaetz, Peter, 222, 226–7
Kalendar of Shephardes, Pynson, 1506, 161
Kamitus, *Treatise of the Pestilence*, Machlinia, 53
Katherine of Aragon, 159
Kay, J., trans. *Siege of Rhodes*, 45
Kele, Thomas, 184
Kempe, Adriaen, 231
Kempe, Margerie, 132
Kendale, John, 43
Kerver, Thielman, 171, 205
Kerver, Thielman, Widow of, 204
Keyser, Martin de, 153
King Apolyn of Tyre, W. de Worde, 1510, 7, 136, 146
King's bookbinder, 181
King's printers, 133, 158, 162, 169, 170, 171, 175, 177, 178, 181
King's stationer, 169
Kinnaird Castle Library, 81
Knight Paris and Fair Vienne, Caxton, 16
Knoblouch, Johann, 211

L., R., bookbinder, 233
Lambertus de Insula, 111
Lambeth Palace Library, 4, 61, 92, 162
Landen, John, 142
Langton, William, 110
Langwyth, Agnes, 177
Lant, Richard, 155, 233
Lauret, Giles, 235
Laurentius of Savona, *Rhetorica Nova*, Caxton, 10
Lauxius, David, 96
Lecomte, Nicholas, 95–97; bindings, 116

INDEX

Leeu, Gerard, 36, 78, 80, 88–91, 215
Lefèvre, R., *History of Jason*, 88
Legenda Francisci, Barbier for Jacobi, 195
Legenda, Sarum, 18
Legrand, J., *Book of good manners*, W. de Worde, 36
Leicester, Earl of, 26
Leland, John, 156
Le Roux, Nicolaus, 204
Le Talleur, G., 55, 57, 59
Lettou, John, 11, 41–44, 130; bindings, 108; with Machlinia, 44–47, 51
Levet, Pierre, 84
Lewis, J., *Life of Caxton*, 39
Liber Assisarum, J. Rastell, 184
Liber Equivocorum, Baligault, 84; Paffroed, 79; Pynson, 63
Liber Festivalis. See Mirk, J.
Liber Synonymorum, Martens, 1493, 79; Hopyl, 1494, 84, 95; Pynson, 1496, 63
Lidgate, J., *Assembly of the Gods*, 15; *Chorle and the Birde*, 10, 16; *Falle of Princes*, Pynson, 1494, 62; *Horse, Shepe, & Ghoos*, Caxton, 10; W. de Worde, 32, 37; *Life of our Lady*, Caxton, 14; *Sege and Destruccyon of Troye*, Pynson, 1513, 163
Life of...Charles the Great, Caxton, 16
Life of Christ, R. Redman, 175
Life of Hyldebrande, W. de Worde, 138
Life of Petronylla, Pynson, 64
Life of St Katherine, W. de Worde, 24
Life of St Margaret, Pynson, 61
Life of St Wenefrede, Caxton, 15
Life of Virgilius, J. van Doesborch, 220, 221
Lily & Erasmus, *De octo orationis partium constructione*, Cambridge, 125
Lily, W., *Grammar*, H. Pepwell, 1539, 149
Lily, W., *Introduction of the Eight parts of Speech*, T. Berthelet, 181
Lincoln Cathedral Library, 49, 132
Linton, W. J., 13
Litill, Clement, 81
Littleton, Sir T., *Tenores Novelli*, Lettou and Machlinia, 44, 46; *Tenures*, Machlinia, 48; Pynson, 57, 173; Redman, 173
London: introduction of printing, 11, 41; bindings, 102
Louvain: printing, 5, 77, 80, 219; binding, 111
Lucianus, *Necromantia*, J. Rastell, 184
Luft, Hans, 228
Lugo, Peregrinus de, *Principia*, Pynson, 1506, 69, 161
Lumley, Lord, 14
Lyndewode, W., *Constitutiones Provinciales*, W. Hopyl, 1506, 194, 197, 205; *Constitutions*, R. Redman, 1534, 176

M., I., border-piece, 176
Maas, Robert, 139
MacCarthy, Count Justin, 73, 74, 162
Macé, Robert, 206
Machlinia, W. de: with Lettou, 44–47; alone, 47–56, 77, 109, 130; bindings, 108
Machyn, Henry, 183
Madan, F., 2, 98
Madden, J. P. A., 95
Magdalen College School, 79
Magna Charta, R. Redman, 1525, 173
Malory, Sir T., *Morte d'Arthur*, Caxton, 16; W. de Worde, 30
Manchester. See John Rylands Library
Mandeville, Sir J., *Travels*, W. de Worde, 1499, 32; Pynson, 64
Manipulus Curatorum, W. de Worde, 1502, 132

Mansion, Colard, 5, 6
Manual, Sarum, B. Rembolt, Paris, 86; Rouen, 1500, 82; Pynson, 1506, 161; C. van Ruremond, 1523, 222, 226; for M. Dotier, 1543, 235
Manual, York, W. de Worde, 1509, 136, 212
Marcant, Nicole, 84
Marchant, John, 204
Marsh Library, Dublin, 143, 234
Marshall, William, 203, 204
Martens, Thierry, 79
Martinus de Predio, 112
Martynson, Simon, 139
Mary of Nemmegen, J. van Doesborch, 220
"Master of St Erasmus," engraver, 142
Maydeston, C., *Directorium sacerdotum*, Caxton, 9; Leeu, 80; Pynson, 70, 71, 159, 161
Maynyal, George, 17
Maynyal, William, 17, 18
Medwall, H., *Interlude of Nature*, W. Rastell, 186
Merry gest...Johan Splynter, J. Notary, 144
Merry jests, J. Rastell, 184
Mery geste of a Sergeaunt and Frere, J. Notary, 145
Meslier, Hugo, 161
Metal engravings, 26, 65, 142
Middleton, William, 124, 125, 176
Miraculous work...at Court of Strete in Kent, 151
Mirk, J., *Liber Festivalis*, Caxton, 14, 105; Hopyl, 96; Morin, 80, 82; Notary, 38; Pynson, 61, 62; Ravynell, 83; W. de Worde, 25, 62, 83
Mirror of Christes Passion, R. Redman, 175
Mirror of Consolation, W. de Worde, 28
Mirror of Golde, 1522, 137, 150
Mirror of the Life of Christ, Pynson, 1503, 161
Mirror of the World, Caxton, 12; L. Andrewe, 140, 156

Mirrour of Our Lady, R. Faques, 1530, 172
Missal, Sarum (?Basle, ab. 1486), 78; Maynyal for Caxton, 1487, 17, 80, 81, 84; M. Morin, 1492, 80, 81; Hertzog for Egmont, 1494, 92, 93; Notary and Barbier, 1498, 38; Pynson, 1500, 68, 159; Higman and Hopyl, 1500, 87; Jean du Pré, 1500, 87, 206; Birckman and Cluen, 1504, 217; Pynson, 1504, 161; Violette, 1509, 207; W. de Worde and R. Faques, 1511, 171; C. van Ruremond, 1527, 223; for W. de Worde and M. de Paule, 207
Missal, York, 1530, 206
Modus tenendi unum hundredum, R. Redman, 174
Montaigne, M. de, 164
Montpellier, Library of Faculty of Medicine, 103
Moore, John, bp of Ely, 8
More, Sir Thomas, 158, 183; *Works*, 1557, 186; *Apology*, 175, 180; *Debellacyon of Salem and Bizance*, 180
Morgan, J. P., 106
Morin, Martin, 80-82, 205-6
Morin, Michael, 206
Morton, Cardinal, 68, 159
Musée Plantin, 80
Music. See *Book of Songs*, 138

N., H., bookbinder, 233
N., I., border-piece, 176
Natura Brevium, R. Redman, 175
Necessary Doctrine and Erudition, 1543, 180
Necton, Robert, 224
Nele, Richard, 193
Newton, Lord, 17
Nicholson, James, 208
Nicholson, John, 225
Nicodemus Gospel, J. Notary, 142; J. Skot, 150-1; W. de Worde, 134
Norwich binding, 108

INDEX

Notary, Julian, 31, 33, 129, 131, 173; at Westminster, 37–40; at London, 141–6; bindings, 119, 145, 232; device, 37–8; method of dating, 135, 141
Nova Festa, Machlinia, 54; Pynson, 61, 65
Nova Rhetorica, St Alban's, 1480, 52
Nova Statuta, Machlinia, 48, 51
Novimagio, Reginaldus de, 74
Nowell, bookbinder, 107, 139, 140
Nut-browne Maide, 151, 215

O., R., bookbinder, 233
Of the newe landes, J. van Doesborch, 220
Offor collection, 39
Oliver, Reginald, 233
Oliver of Castile, W. de Worde, 1518, 137
Olivier, Petrus, 82, 205
Orchard of Syon, W. de Worde, 1519, 137
Ordinale, Sarum, Caxton, 9, 22
Ordynaunce...Kynge'sEschequier, Middleton, 124
Origen, *De beata Maria Magdalena*, W. Faques, 170
Ortus Vocabulorum, 194, 197
Os, Govaert van, 30, 33
Osborne, Thomas, 9
Osterley Park Library, 16
Oswen, John, 228
Ovidius, *Metamorphoses*, 14
Owen, David, 193
Oxford libraries:
  Bodleian, 10, 21, 25, 28, 58, 59, 61, 68, 81, 82, 83, 90, 95, 106, 108, 112, 132, 153, 154, 180, 198, 199, 210, 212, 216, 231, 232
  Brasenose College, 80
  Corpus Christi College, 49, 92, 112, 115, 139
  Merton College, 8
  New College, 17, 125, 196
  St John's College, 160, 200
Oxford, printing, 1, 2, 41, 79, 187; booksellers, 69, 77, 98, 161, 196; bindings, 103, 104

Paffroed, R., 79
Palsgrave, J., *L'eclarcissement de la langue Française*, 166
Panzer, G. W., 217
Paper-making in England, 29
Paris printing, 17, 18, 84, 85, 87, 96, 97, 161, 193, 194, 195, 196, 199, 205, 210, 214
Paris and Vienne, G. Leeu, 88
Parker, H., *Dives and Pauper*, Pynson, 1493, 54, 61
Parliament of Devils, W. de Worde, 135
Parron, William, 68
Parson of Kalenborowe, J. van Doesborch, 220, 221
Parvula, N. Marcant, 84
Parvulorum institutio, J. Butler, 1529, 152
Passio Jesu Christi, Strasburg, 1506, 211
Passion of our Lord (ab. 1508), 211
Pastime of People, J. Rastell, 184
Paule, Michael de, 207
Paynell, T., *Assault and Conquest of Heaven*, 1529, 178; trans. of *Regimen sanitatis Salerni*, T. Berthelet, 177, 178
Pelgrim, Joyce, 193–198, 232
Penketh, Thomas, 42
Pepwell, Arthur, 150
Pepwell, Henry, 139, 147–150, 199, 202
Pepwell, Ursula, 149
Pepysian Library. *See* Cambridge libraries
Perez de Valentia, J., 42
Perott, N., *Regulae Grammaticales*, Louvain, 77
Peter Post Pascha, 66, 67, 93
Peterborough Cathedral Library, 132
Petyt, Thomas, 176
Pigouchet, Philippe, 115, 205
Pilgrimage of Sir Richard Guylforde, Pynson, 1511, 163

Play concerning Lucretia, J. Rastell, 185
Poitevin, Jean, 86
Pollard, John, 235
Pomander of Prayer, R. Redman, 175
Powell, Thomas, 181, 183
Prevost, Nicolas, 199, 205
Prices of paper and printing, 169, 179
Processional, Sarum, Pynson, 1502, 160; N. Prevost, 1530, 218; C. van Ruremond, 1523, 222, 225–6; Antwerp, for J. Reynes, 200
Proclamation on the coinage, W. Faques, 1504, 169
Proclamations, T. Berthelet, 179
Proctor, R., 17, 125, 196
Prognostications, 68, 226–7
Promise of Matrimony, Machlinia, 48, 51
Promptorium parvulorum, 67
Promptorius puerorum, Pynson, 1499, 66, 93
Propositio Johannis Russell, Caxton, 4, 9
Prymer, J. Byddell, 1535, 204
Psalter, Latin, Caxton, 13; W. Faques, 1504, 169, 170
Psalter, Sarum, C. van Ruremond, 226
Psalterium beate marie virginis, S. Voster, 210
Psalterium cum hymnis, 1507, 194; 1530, 199
Pudsey, Bishop, 102
Pynson, Margaret, 166
Pynson, R., 55–71, 158–169, 173, 174, 177, 189, 206, 235, 236
  Arms, 162
  Bindings, 109, 113–115, 165, 232
  Borders, 49, 56, 176
  Books printed by, 93, 124, 145, 195
  Devices, 59, 62, 65, 174
  King's printer, 133, 162, 171
  Method of dating, 68, 159, 160
  Printing office, 40, 71, 141, 158, 159, 173

Pynson, R. (*continued*):
  Types, 58, 60, 61, 62, 70
  Woodcuts, 57, 62, 64
Pynson, R., junior, 166, 177

Quarto leaves in a folio volume, 52
Quatre derrenières choses, Caxton, 6
Quattuor Sermones. See Mirk, J., *Liber Festivalis*
Quentell, H., 65, 79

R., A., bookbinder, 116
R., G., bookbinder, 233
R., P., device, 83
Raimund, John. *See* Ruremond, Hans van
Rastell, Elizabeth, 183
Rastell, John, 152, 156, 158, 183–186, 203; *New Boke of Purgatory*, 184
Rastell, John, junior, 185
Rastell, William, 158, 180, 185, 186
Ratcliffe, John, 79, 162
Ravynell, James, 83
Rawlinson Collection, 134
Recuyell of the Historyes of Troye, Caxton, 3, 6
Redman, Elizabeth, 176
Redman, John, 125, 177
Redman, Robert, 155, 158, 165, 172–177, 203
Regnault, Francis, 207–209
Regula Benedicti, W. de Worde for Jacobi, 196
Regulae et ordinationes, Machlinia, 55
Rembolt, Berthold, 87, 205
Remonde. *See* Ruremond
Revelation of St Nicholas, Machlinia, 48, 50
Reynard the Fox, Caxton, 13; W. de Worde, 37
Reynes, John, 119, 193, 199–202, 219; bindings, 232–234
Reynes, Lucy, 201, 202
Richard, Jean, 81, 82, 205, 207
Richard Cœur de Lion, W. de Worde, 1509, 135

Richardson, John, 235
Richmond, Countess of, 133, 134
Ripon Cathedral Library, 70, 159
Rivers, Earl, 6
Robin Hood, J. van Doesborch, 220
Roce, Denis, 150
Rolle, Richard. *See* Hampole, Richard de
Rome, 41
Rood, Theodore, 104
Rosary, J. Skot, 1537, 151
Rosse, Denis, 150
Rote or mirror of consolation, W. de Worde, 33, 34
Rouen: printing, 37, 55, 80, 82, 83, 204–6, 214; bindings, 233
Rouse, John, 235
Row, John, bookbinder, 224, 235
Roxburghe sale, 35, 39
Royal Book, Caxton, 16, 106
Rue, Andrew, 193
Rue, John, 193
Rule of St Benet, Caxton, 21; Pynson, 160
Rupertus, *De victoria verbi Dei*, 1541, 219
Ruremond, C. van, 130, 200, 222–227, 230; Widow of, 230
Ruremond, Hans van, 218, 222–7
Russell, J., bp of Lincoln, 4

St Alban's, printing, 41, 149, 187; binding, 109
St Alban's Grammar School Library, 106, 107
St Andrews printing, 151
Saint Germain, C., *Answer*, 157; *Division*, 175, 180; *Dialogue*, 180
St Paul's Churchyard, 191
Savonarola, *Sermo*, R. Pynson, 1509, 163; Tracts, H. Jacobi, 1510, 198
Sayle, C. E., 228, 239
Scala Perfectionis, J. Notary, 1508, 143
"Scales" binder, 110
Schoeffer, J., 229
Schoiffer, P., 4

Scotland, introduction of printing, 7
Scott, E. J. L., 130
Scriptores rei rusticae, Reggio, 1496, 80
Sedley, John, 194
Selden, John, 106
Sermo pro episcopo puerorum, W. de Worde, 28
Sermones dormi secure, 193
Service books printed abroad for the English market, 78, 80–82, 84–87, 91–93, 205–210, 212–213, 222, 225, 226
Seven Points of True Love, Caxton, 21
Seven wise masters of Rome, W. de Worde, 36
Sex quam elegantissimae epistolae, Caxton, 14
Shepherdes Calendar, 145
Shirburn Castle Library, 184
Short treatyse (Margerie Kempe), W. de Worde, 132
Shrewsbury School Library, 133
Siberch, John, 99, 130, 226
Siege of Rhodes, 45
Signatures first used, 3; in England, 11
Signs:
A.B.C., R. Faques, 171
Fat Hen, Birckman, 219
George, Pynson, 71, 159, 173; R. Redman, 173
Golden Cross, L. Andrewe, 155
Golden Missal, A. van Berghen, 215
Great Golden Mortar, A. van Berghen, 215
Iron Balance, R. van den Dorp, J. van Doesborch, 220
Lucrece, T. Berthelet, 178
Maiden's Head, R. Faques, 171, 234
Mermaid, J. Rastell, 144, 184
Our Lady of Pity, W. de Worde, 135; J. Byddell, 144; J. Redman, 177
Red Pale, Caxton, 6

Signs (*continued*):
  Rose Garland, R. Copland, 146, 147
  St Anne, J. Pelgrim, 195
  St John Evangelist, G. Chasteleyn, 69; J. Butler, 152; R. Wyer, 156
  St Katherine, 198
  St Mark, J. Notary, 144
  St Nicholas, N. Lecomte, 95; J. Toy, 153
  Striped Ass, P. de Couvelance, 198, 199
  Sun, W. de Worde, 33, 131, 146; J. Byddell, 140, 144; J. Gaver, 140
  Three Kings, J. Notary, 141, 143, 144
  Trinity, 198, 199; H. Pepwell, 147; H. Jacobi, 148, 195; H. Smith, 176
  Wodows, P. Treveris, 115
Silverlink, John, 218
Singer, S. W., 2
Sirectus, Antonius, *Formalitates*, 196
Sizes of books, 49
Skelton, John, 64; *Bowge of Court*, W. de Worde, 37; *Magnificence*, J. Rastell, 185
Skot, John, 137, 138, 150–53
Smarte, *Epitaph of Jasper, Duke of Bedford*, Pynson, 63
Smith, Henry, 176
Smith, Richard, 142, 149
Smyth, Thomas, 154, 155
Snowe, John, 166
Society of Antiquaries Library, 55, 102, 155
Solomon and Marcolphus, G. Leeu, 88, 89
Southwark printing, 177, 199, 208
Speculum Christiani, Machlinia, 53, 77
Speculum Spiritualium, W. Hopyl, 1510, 194
Speculum vitae Christi. See Bonaventura
Spencer, Earl, 8, 16, 74. See *also* Althorp Library

Spering, Nicolas, 117, 120, 216
Spiritus Guidonis (Pynson), 59
Squire, Henry, 158, 189
Stanbridge, John, 79, 206; *Accidence*, J. Gaver, 140; J. Skot, 150; *Shorter Accidence*, 1534, 153
Stans Puer ad Mensam, 10
Statham, N., *Abridgement*, Rouen, for Pynson, 57
Stationer, business of, 72, 76
Stationer to the King, 1485, 77
Stationer, University, at Oxford, 77
Stationers: foreign, in England, 54; fifteenth century, 72–100; of London, 187–213; aliens, 187–191; from Antwerp, 214–231
Stationers' Company, 76, 94, 95, 148, 188, 189, 227
Statutes first written in English, 56
Statutes, Machlinia, 54; W. Faques, 169, 170; T. Berthelet, 178
Statutes of War, Pynson, 1513, 163
Stewarde, William, 141
Stewart, W., bp of Aberdeen, 124
Stillingfleet, Bishop, 143
Stoke, Thomas, 235
Stokeslay, Bishop, 148, 149
Stondo, Bernard van, 53, 77
Stonyhurst Library, 32, 161
Storys and prophesis, S. Cock, 1536, 231
Strasburg printing, 211
Strype, John, 191
Sturbridge fair, 192
Suethon, Ludovicus, 199, 202
Sulpitius, J., *Grammar*, W. de Worde, 1504, 133
Sutherland, Duke of, 38
Sutton, Edward, 201
Sutton, Lewis, 148, 199, 202–3
Sutton, Nicholas, 154, 203
Sykes sale, 40
Symmen, Henric van, 90
Symonds, Thomas, 192, 202

Tab, Henry, 149
Tanner, Thomas, 11, 90
Tate, J., papermaker, 29
Taverner, John, 193, 194
Taverner, Richard, 154
Temple of Brass, Caxton, 10
Temple of Glas, Caxton, 10
Terentius, Pynson, 66; Paris, 1504, 206; *Hecyra*, Pynson, 1495, 63; *Vulgaria*, Machlinia, 48, 51, 54; G. Leeu, 78; Oxford, 79; W. Faques, 170
Three Kings of Cologne, W. de Worde, 28, 34, 90
Thwaytes, Edward, 151
Thynne, William, 156
Tindale, W., 228, 229, 230
Title-page, first English, 53; first Westminster, 25
Title-pages, Collection of, in British Museum, 8
Toy, John, 150, 153, 154, 203
Treatise of Love, W. de Worde, 24, 25
Tree and XII frutes of the holy goost, 172
Treveris, Peter, 115, 156, 199, 216
Trevisa, John, 29
Trinity stationers. See Jacobi, Henry; Pelgrim, Joyce
Triphook, bookseller, 53
Troost, Ian, 228
Troylus and Cressede, W. de Worde, 1517, 137
Tuke, Sir Brian, 156
Tunstall, C., *De arte supputandi*, Pynson, 1522, 165
Twelve Profits of Tribulation, Caxton, 21
Tyl Howleglas. See Howleglas
Type, side impression of, 216

Ulmer, Johann, 219
Upsala University Library, 11
Utrecht printing, 5, 217

Valuation of gold and silver, J. van Doesborch, 155, 221
Vanduffle, Woter, 112

Vatican Library, 164
Vaughan, Stephen, 168
Veldener, J., 5, 52, 54
Vellum, books printed on, 20, 30, 68, 73, 125, 134, 136, 137, 150, 160–5, 170, 180, 181, 204, 209
Venice printing, 73, 85, 91, 93
Verard, Antoine, 145, 161, 206, 212
Vergilius, *Eneydos*, Caxton, 20; *Life of*, J. van Doesborch, 220, 221
Very declaration of the...will of man, St Alban's, 149
Violette, Pierre, 205, 206, 207
Vitas Patrum, W. de Worde, 1495, 28
Voragine, J. de, *Golden Legend*, Caxton, 15, 18, 23, 24; W. de Worde, 24, 25, 30, 31, 135, 141; Notary, 31, 141; Pynson, 162
Voster, Simon, 210
Vrankenbergh, H., 53

Wakefield, R., *Kotser codicis*, T. Berthelet, 178
Wallensis, T., *Expositiones super Psalterium*, Lettou, 42, 43, 108
Wallis, Thomas, 193
Walsingham, *Foundation of Our Lady's Chapel*, Pynson, 63, 64
Wanseford, Gerard, 94
Warde, son-in-law to Pynson, 166
Watson, Henry, 135, 136
Watton, John, 53
Way, Albert, 67
Way to the Holy Land, W. de Worde, 137
Waynflete, William, 79
Weale, W. H. J., 111
Wednesday's Fast, W. de Worde, 8
Wenssler, Michael, 78
Westminster Abbey Library, 68, 99, 103, 113, 125, 148, 212, 226
Westminster printing, 1–40, 187
Whitchurch, Edward, 208, 209
Whitford, R., *Dayly Exercise*, R. Redman, 175

INDEX

Whitinton, R., *Grammar*, W. de Worde, 124, 137; Pynson, 1515, 62; *Vulgaria*, Pynson, 1520, 177
Wilcock, William, 42, 45
Willett sale, 61
Winchester bindings, 102–104
Winchester Domesday Book, 102
Windsor, Royal Library, 20, 81
Wislyn, John, 139
Withers, Richard, 166
Wolsey, Cardinal, 164, 224
Wonderful shape and nature, J. van Doesborch, 155, 221
Wood, Anthony á, 178
Woodcut of Crucifixion, 21, 27, 33, 132–3
Woodcuts, 12, 57, 62, 64, 86, 88, 89, 91, 135, 141, 154, 156, 162, 172, 184, 194, 199, 211
Worde, Elizabeth de, 130, 139
Worde, W. de, 23–38, 56, 129–141, 146–7, 149, 150, 152, 162, 178, 199, 235
  Bindings, 107, 108, 113, 139, 140, 197
  Books printed by, 4, 7, 8, 90, 124, 145, 196, 207, 212

Worde, W. de (*continued*):
  Death, 138
  Devices, 25, 26, 32, 132, 133, 212
  Method of dating, 31, 135, 141
  Number of books printed, 34, 137
  Printer to the King's mother, 133–4
  Printing office and shop, 33, 131, 135
  Quarto tracts, 32
  Removal to London, 33, 131
  Types, 19, 24, 25, 26, 30, 131
  Woodcuts, 26, 27, 28, 29, 30, 32, 33, 36, 37, 131, 133, 135, 142
Wright, Edward, 201
Wyer, Robert, 156, 203

Year-books, Lettou and Machlinia, 44; Pynson, 57, 59
Yonge, John, 198
York, printing, 87; stationer, 94

Zel, Ulric, 142